T0343300

Solid Waste Recycling and Processing

Planning of Solid Waste Recycling Facilites and Programs

Solid Waste Recycling and Processing

Planning of Solid Waste Recycling Facilities and Programs

Second Edition

Marc J. Rogoff, PhD

AMSTERDAM • BOSTON • HEIDELBERG • LONDON • NEW YORK
OXFORD • PARIS • SAN DIEGO • SAN FRANCISCO • SINGAPORE
SYDNEY • TOKYO
William Andrew is an imprint of Elsevier

William Andrew is an imprint of Elsevier
225 Wyman Street, Waltham, 02451, USA
The Boulevard, Langford Lane, Kidlington, Oxford OX5 1GB, UK

1st edition published 1994 with the title Approaches to Implementing Solid Waste Recycling Facilities

Library of Congress Cataloging-in-Publication Data
A catalog record for this book is available from the Library of Congress

British Library Cataloguing-in-Publication Data
A catalogue record for this book is available from the British Library

ISBN: 978-1-4557-3192-3

For information on all Elsevier publications visit our web site at elsevierdirect.com

Printed and bound by CPI Group (UK) Ltd, Croydon, CR0 4YY

Contents

About the Author

Marc J. Rogoff is currently a Project Director with SCS Engineers. He has over 30 years of experience in solid waste management as a public agency manager and consultant. He has managed more than 200 consulting assignments across the world on literally all facets of solid waste management, including waste collection studies, facility feasibility assessments, facility site selection, property acquisition, environmental permitting, operation plan development, solid waste facility benchmarking, ordinance development, solid waste plans, financial assessments, rate studies/audits, development of construction procurement documents, bid and RFP evaluation, contract negotiation, and bond financings.

Dr. Rogoff has directed engineer's feasibility reports for nearly two dozen public works projects totaling more than $1.2 billion in project financings. He has interacted with bond rating agencies, financial advisors, insurance underwriters, and investment bankers involved in these financings. His efforts have included the development of detailed spreadsheet rate models establishing the financial feasibility of each project, long-term economic forecasts, and projected rate impact upon project users and customers.

He has been professionally active during his career serving as Chair of the Solid Waste Technical Committee for the American Public Works Association and Chair of the Collection and Transfer Technical Division of the Solid Waste Association of North America (SWANA). He has published widely in a variety of technical trade publications on solid waste management and recycling as well as presented technical presentations at international, national, and regional symposiums and conferences. He has authored or coauthored five major textbooks dealing with solid waste management.

Preface

This revision of the book builds on major changes that have occurred in the recycling industry over the past 20 years. For example, most states have implemented mandatory recycling goals and many are well into their third and fourth phase to raise their initial goals, some exceeding 75% recycling, and a few regions even have made a goal of "zero waste."

Furthermore, the formal recycling of construction and demolition debris has matured in many regions with some companies operating multiple, semi-automated sorting facilities that are as sophisticated as any solid waste facility. The introduction of alternative waste conversion facilities not only in the United States, but globally, has called for stepped-up processing schemes. These technologies, utilizing advanced thermal, chemical, and biological processes, can require extensive preprocessing of feedstock to produce precise size fraction and consistent quality to maximize energy production.

During our voyage, I take an unvarnished look and provide a critical eye at what has worked and what has not over this period, and assess the next generation of challenges and possible solutions. For example, there is a continuing debate within the solid waste industry about the merits of curbside sort versus single-stream recycling and the use of advanced technology in materials recovery facilities. For those of you who are newcomers to our profession, I have provided an extensive set of definitions and explanations of the common-day acronyms. My overall goal is to provide concise summaries of emerging concepts in recycling worldwide such as zero waste, sustainability, LEED certification, and means to pay and finance these systems.

Lastly, I have provided a series of "case studies" to fill out our discussion. I have collected examples from the "best-in-class" communities, agencies, and organizations, which highlight, in my opinion, best practices in solid waste management and recycling. My hope is that this information will enable your community or organization to become "best-in-class."

Happy reading!

Marc J. Rogoff
Tampa, FL, USA

Acknowledgments

I am deeply grateful to the support of my family during the countless hours of research and writing to help develop this new edition. I would never had completed this project without their understanding.

This "project" began while I was employed at SCS Engineers, where I have worked closely with many colleagues discussing the topics that are outlined in this book. Specifically, I would note the assistance and support of my friends at SCS: Steve Anderson, Bruce Clark, Bob Gardner, Michelle Leonard, Sarah Norton, and David Ross.

I would also like to acknowledge the support of numerous friends and colleagues at the Solid Waste Association of North America, Keith Howard, Roger Lescrynski, Barry Shanoff, Phillip Stecker, who provided background information about their projects as well as excellent photographs to help illustrate the text. Without the assistance of these folks, this book would not have been written.

The author would also like to thank the following recycling facility program operators who contributed their time in discussing their programs, critiquing initial reviews of these discussions for the program fact sheets, and kindly providing useful photos depicting facets of their programs: Kim Brunson, Michael Bush, Corey Hawkey, Alex Helou, Helen Howes, Elizabeth Leavitt, Jack Macy, Lisa Moore, Vincent Sferrazza, and Brett Stave. I would like to especially thank Helena Bergman for her assistance in providing research information on recycling activities on various European cities.

Finally, thanks to my publishing team from Elsevier Inc., including David Jackson and Colin Williams.

Marc J. Rogoff

Acronyms and Abbreviations

A/E	Architect Engineer
API	American Paper Institute
ASTSWO	Association of State and Territorial Solid Waste Management Officials
Btu	British thermal unit
CCR	California Climate Registry
CFR	Code of Federal Regulations
CPRR	Center for Plastics Recycling Research
CRT	Cathode ray tube
e.g.	For example (exempli. gratia)
EPA	US Environmental Protection Agency
et al.	And others (et alia)
E-Waste	Electronic waste
ft.	Foot (feet)
FY	Fiscal year
GHG	Greenhouse gases
HDPE	High density polyethylene
HHW	Household hazardous waste
HRA	Health risk assessment
i.e.	That is (id est)
IDB	Industrial Development Bond
Lb	Pound
LDPE	Low density polyethylene
MRF	Materials recovery facility
MSW	Municipal solid waste
MTCO2E	Metric tons carbon dioxide equivalence
MWC	Municipal waste combustor
NA	Not available
NIMBY	Not-in-my-backyard
No.	Number
NMOC	Nonmethane organic compounds
OCC	Old corrugated cardboard
ONP	Old newspaper
O & M	Operations and maintenance
OWP	Office waste paper
p.	Page
pp.	Pages
PCD	Pounds per capita per day
PAYT	Pay as you throw
PET	Polyethylene terephthalate
PP	Polypropylene

PS	Polystyrene
PVC	Polyvinyl chloride
RCRA	Resource Conservation and Recovery Act
RFP	Request-for-proposal
RFQ	Request-for-qualifications
RMP	Residential mixed paper
SMART	Save money and reduce trash
SSR	Source separated recyclables
SWANA	Solid Waste Association of North America
T	Ton(s)
Tpd	Tons per day
Tph	Tons per hour
Tpy	Tons per year
US	United States
VOCs	Volatile organic compounds
WARM	Waste reduction model
WTE	Waste to energy
Yd	Yard
Yr	Year

Signs and Symbols

>	Greater than
<	Less than
#	Number
“	Inches
$	Dollars

1 Introduction

When the first edition of this book was written more than 20 years ago, the modern ways of recycling in North America was in its infancy. At that time, the US Environmental Protection Agency (EPA) and various state and provincial governments were implementing solid waste plans mandating some form of local solid waste agency recycling program, these agencies had just began offering recycling opportunities, and private vendors were just constructing the first recycling processing facilities.

Over the last 20 years, the recycling industry worldwide has made remarkable advances in the forms of programs offered to the public and the types of technologies deployed to process recyclable commodities for market. Solid waste facilities have been upgraded, and once formerly contaminated sites have been cleaned up. According to the best estimates available, there are currently about 1900 landfills in the United States, dropping significantly from about 8000 in 1988 with the remaining landfills much larger and serving regional needs [1]. Recycling programs have become "mainstream" with most local solid waste agencies across the United States. As of 2010, there are an estimated 9000 curbside collection programs in the United States [1]. Yet, as Figure 1.1 graphically illustrates, while the number of recycling programs have dramatically increased in recent decades, more work needs to be done in the United States to increase the role of recycling in diverting materials from landfills and waste combustion sources.

Current Waste Generation

United States

The United States generates more waste per person than any nation in the world. The best available source of data on waste generation is provided by the EPA as part of its continuing updates of studies conducted by the agency during the 1960s to estimate material flows in the national economy. Currently, these studies estimate that the United States is generating roughly about 250 million tons of municipal solid waste (MSW) per year, at 4.6 pounds per person per year, including residential, commercial, and institutional sources. Figure 1.2 illustrates estimated waste generation rates from 1960 to 2010, the last update released by EPA. It is noteworthy to point out that the data shows that while the total MSW has increased over this period, the per capita rate of waste generation has remained relatively stable since the 1990s, averaging about a 3% annual increase.

While the generation of MSW has increased substantially from 3.66 to 4.43 pounds per person per year from 1980 to 2010, the recycling rate has increased as well from an estimated 10% in 1980 to 34% in 2010. Disposal of MSW in landfills has decreased from an estimated 89% in 1980 to about 54% in 2010.

Solid Waste Recycling and Processing. DOI: http://dx.doi.org/10.1016/B978-1-4557-3192-3.00001-4

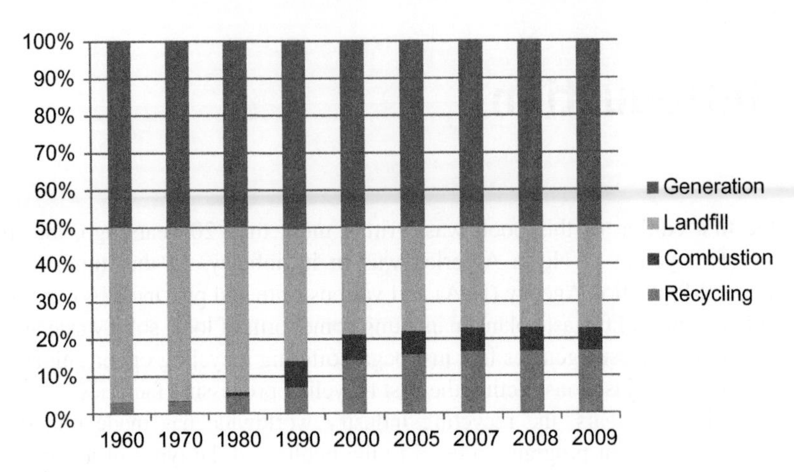

Figure 1.1 Solid waste trends in the United States, 1960–2009. (From Ref. [1])

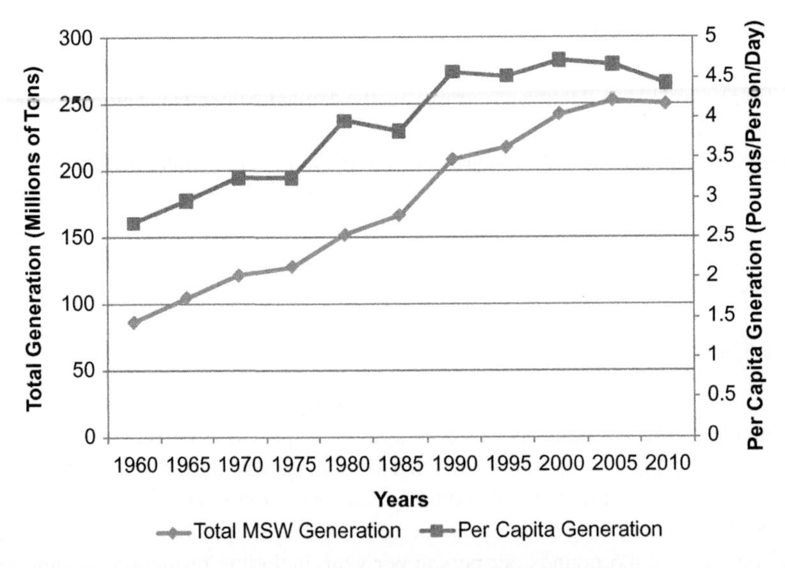

Figure 1.2 MSW generation rates in the United States. (From Ref. [1])

Figure 1.3 illustrates the material types that are projected by EPA to make up the national MSW waste stream. Organics such as paper, paperboard, yard waste, and food scraps make up the largest component of the national waste stream at 56%.

Figure 1.4 illustrates the estimated recycling rates of selected products within the MSW stream. According to the EPA, newspaper and associated paper recovery was about 72% in 2010 with about 58% of yard wastes were recovered. Metal

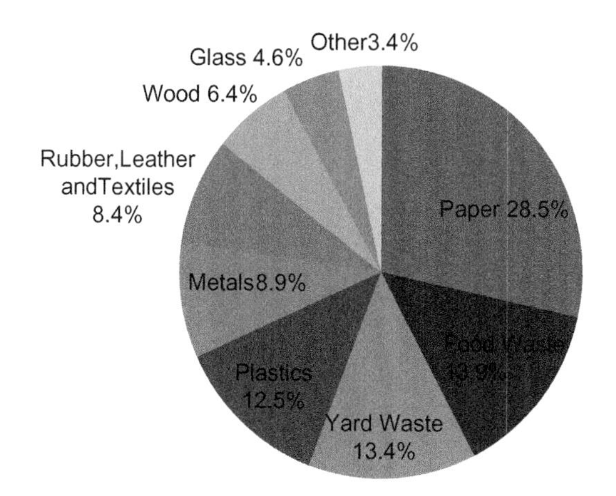

Figure 1.3 Total US MSW generation by category, 2010. (From Ref. [1])

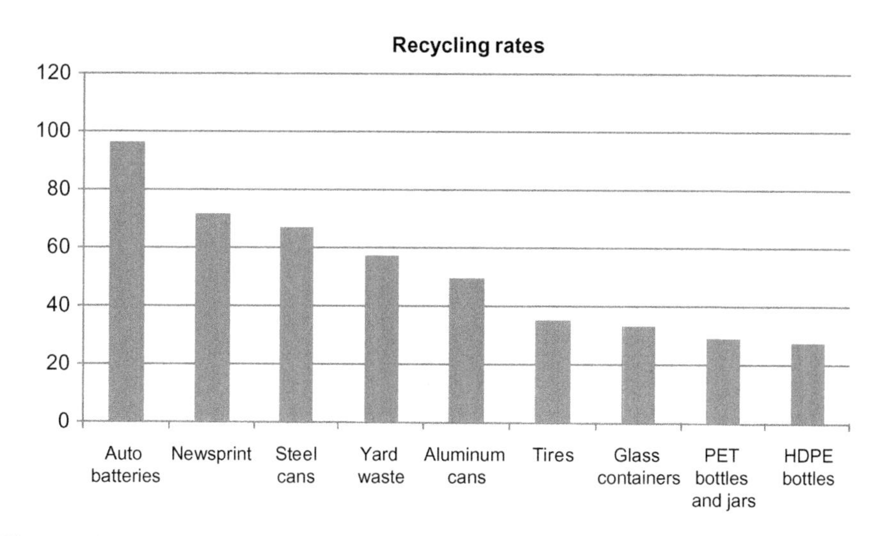

Figure 1.4 Recycling rates of selected products in the United States, 2010. (From Ref. [1])

cans (ferrous and aluminum) were recycled in the United States at about 50%. Other materials such as tires, glass bottles, and plastics still have a significantly low recovery rate. With 96% recycling rate, lead-acid batteries continue to remain as one of the most highly recycled products.

There are limited current data that compares worldwide waste generation rates, albeit only those representing industrialized countries. The most complete data set has been compiled by the Conference Board of Canada utilizing information from countries deemed "high income" by the World Bank [2]. The intent of the

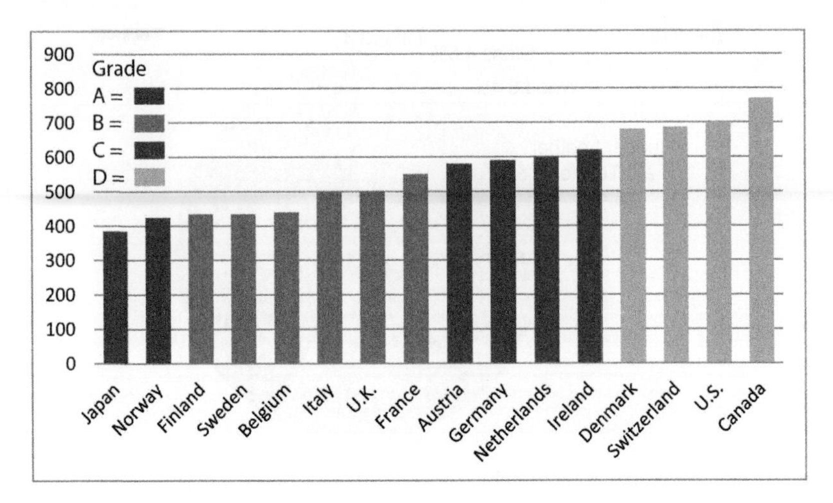

Figure 1.5 Comparison of municipal waste generation. (From Ref. [2])

Conference Board's research study was to develop a benchmarking database comparing countries similar in population, land mass, and income to Canada. Figure 1.5 shows the recycling rates of 16 "peer countries." Further, the Conference Board assigned a report card style ranking (A, B, C, and D) using the following methodology:

- The difference between the top and the bottom recycling performers was calculated and this figure was divided by four.
- A country received a report card rating of "A" if its score is in the top quartile; a "B" if its score is in the second quartile; a "C" if its score is in the third quartile; and a "D" if its score is in the bottom quartile.

For example, the top performer, Japan, ranks first out of the 16 peer countries and scores an "A" grade on its recycling score card. In comparison, according to the Conference Board's findings, Canada ranks in last place and receives a "D" grade on the municipal waste generation report card. Canada produces 777 kg per capita, twice as much as Japan in 2009, the most recent data available.

The New Solid Waste Paradigm—"Materials Management"

In recent years, many solid waste practitioners and agencies have argued for a new approach to replace the "end of pipe" waste management technologies such as landfilling and waste incineration. This is commonly termed the "materials management" paradigm, which aims at a comprehensive evaluation of how materials are managed upstream of traditional waste management techniques so they can be sustainably managed at all stages of their life cycle throughout the economy (Figure 1.6). That is, this new paradigm helps address all the stages of materials and products from raw

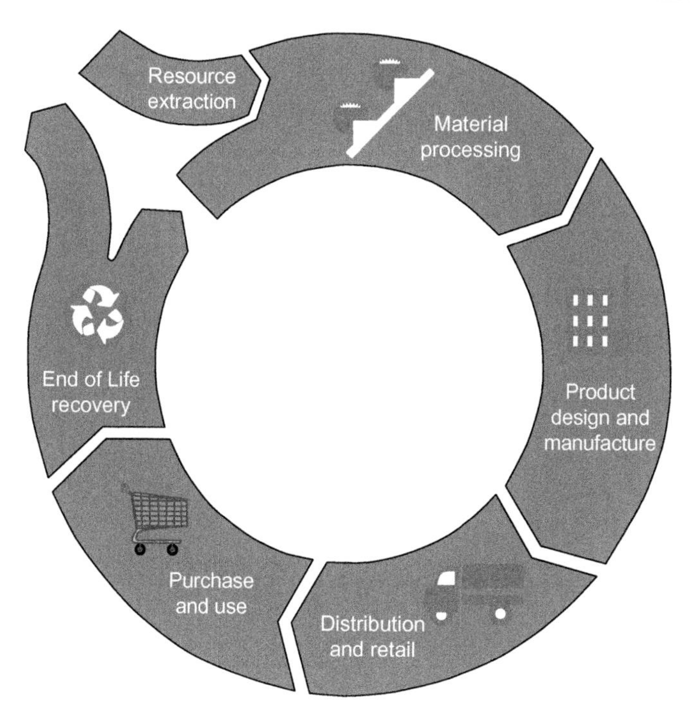

Figure 1.6 Life cycle of materials and products. (Adapted from Ref. [3])

material extraction to product design, manufacture, and transport, consumption by the consumer, use, reuse, recycling, and then final disposal.

In essence, this new paradigm puts the highest value on source reduction and extended producer responsibility and then efforts for waste conversion such as energy, biofuels, and compost. In comparison to the traditional waste hierarchy pyramid, which has shown landfills serving as the base of the pyramid, these waste disposal technologies are considered as an option of "last resort" by many in the solid waste industry for those materials that cannot be recovered for beneficial reuse [4]. This new paradigm is broadly based on visions of future solid waste management expressed by both the EPA and the European Union. This theme has been adopted by many solid waste agencies as they consider revisions to their existing solid waste management plans. We will provide some illustrative examples of these efforts in Chapter 2.

Long-Term International "Drivers"

As this book is written, an upturn in the world's economies is being felt from the significant declines experienced by most economies during the "Great Recession." Notwithstanding the remaining economic consequences, there are five major trends, in the author's opinion, that will greatly influence the move toward increased recycling worldwide.

Energy

Yergin [5], in his classic review of the oil industry, "The Prize," points out that there have been many revelations by industry observers that the world was running out of petroleum and, as a consequence, steep increase in the price of oil would soon follow. The end of the era of "cheap oil" has been predicted many times since the major oil fields were first discovered in the 1860s in Baku, Russia near the Black Sea. Time and again, new fields were discovered with new technology (e.g., offshore drilling, deep wells, remote locations, fracking, and shale oil) only to confound the predictions of a sudden spike in prices and capacity.

Nevertheless, it is unlikely to assume that new energy sources will enable us to assume that no supply interruptions will occur or energy prices will not increase. To this observer, sudden spikes in energy prices will become more frequent and volatile because the world is so much more integrated than ever. A major revolution in an oil producer or a catastrophic well blowout in one part of the globe appears to impact energy prices almost overnight. This volatility impacts all that we do in solid waste management from collection, recycling, and processing to disposal technologies. While we have retreated from triple digit prices for petroleum as of this writing, most energy experts agree that the era of cheap energy is over.

Climate Change

Climate change is commonly defined by atmospheric scientists as a significant and long-lasting change of weather patterns over periods of decades to millions of years. There is much debate in some political circles and business organizations that these changes are primarily caused by biotic sources, variations in solar radiation, volcanic eruptions, and ocean circulation. Some observers have opined that climate changes have occurred in the past and is not associated with man's combustion of fossil fuels. Scientific consensus, however, is "that climate is changing and that these changes are in large part caused by human activities" [6]. While much remains to be explained, the overwhelming evidence suggests that hypotheses and scientific models of climate change have stood firm in light of challenging scientific debate over the past few decades [7]. The release of carbon dioxide in the atmosphere since the beginning of the Industrial Era has resulted in a rapid increase in global temperatures, a reduction of ice in the polar caps, a rise in sea levels, and more dynamic weather extremes. Climate change is expected to impact the hardest—changes in precipitation levels, rising sea levels, temperatures that impact crop production, and weather-related events.

What does the trend in climate change mean to solid waste management? The real answer in a nutshell is the assumed relationship of solid waste management activities and greenhouse gas (GHG) emissions (Table 1.1). The best available scientific evidence suggests that atmospheric concentrations of CO_2 must be stabilized, requiring a reduction of GHGs.

Table 1.1 GHGs from Common Solid Waste Management Activities

Waste Management Activity	Potential GHG Emissions
Collection (recyclables and mixed waste)	Combustion of diesel in collection vehicles Production of diesel and electricity (used in garage)
Material recovery facilities	Combustion of diesel used in rolling stock (front-end loaders, etc.) Production of diesel and electricity (used in building and for equipment)
Yard waste composting facility	Combustion of diesel used in rolling stock Production of diesel and electricity (used for equipment)
Combustion (also referred to as waste to energy)	Combustion of waste Offsets from electricity produced
Landfill	Decomposition of waste Combustion of diesel used in rolling stock Production of diesel Offsets from electricity or steam produced
Transportation	Combustion of diesel used in vehicles Production of diesel
Reprocessing of recyclables	Offsets (net gains or decreases) from reprocessing recyclables recovered Offsets include energy- and process-related data

Source: From Ref. [8]

Natural Resource Depletion

Over the last few decades, the economies of China, India, Brazil, South Korea, and other countries have expanded rapidly producing a consumer economy similar to the United States, Europe, and Japan. The material aspirations of populations in these developing economies pose increasing concerns about sustainability of the global environment. This has increased the demand for increasingly scarce energy resources, timber, agricultural commodities, and minerals [9]. This has caused a rather substantial upswing of long-term trade deals in Africa and South America to secure these resources for these economies. Pressures on these increasingly scarce natural resources will continue as worldwide demand expands resulting in changes to consumption patterns and trends to substitute these materials through increased recycling and substitution.

Increasing Commodity Prices

The global economy is becoming more highly integrated than ever in recorded human history. This has resulted from a loosening of trade barriers enabling in a freer movement of money and materials and a higher prosperity in many countries. The integration of world economies poses unprecedented opportunities for goods and services, and an increasing demand for manufacturing capacity and minerals, which most are finite, will tend to be more dispersed across the globe. Undoubtedly, there are pluses and minuses to these global trends for sustainable development.

Information Revolution

Through the development of increasingly integrated communication resources, the amounts and variety of information will become available to nearly everyone on the planet. Through the use of the Internet, cell phones, and social media applications enables the ability for billions to access and share this information in a timely fashion. In a moment's notice, breaking developments and news on one side of the globe can be shared. It is hard to underestimate the impact the information revolution will have on politics and the way business is conducted in many nations.

What does this mean to solid waste management and recycling in general? In my opinion, a more efficient dissemination of information about the workings of recycling markets and consumer choices will translate into more sustainable solutions for both consumers and businesses. Greater awareness of environmental trade-offs should result in more choices for products and services.

Increased Corporate and Governmental Sustainability

The increasing cost of energy and materials and the impacts of global pollution are being recognized by business as significant threats to their competiveness and overall profitability. However, all of these external threats also pose opportunities for corporations to find ways to reduce costs, become more efficient, and, at the same time, become less polluting and more sustainable. In their groundbreaking book, *Cradle to Cradle: Remaking the Way We Make Things*, which was published in 2002, architect Bill McDonough and the coauthor, Michael Braungart, influenced the current generation of industrial designers and corporate sustainability officers by arguing that they must radically alter the designs of their products to make them more sustainable [10]. A recent sequel to this seminal book in 2013, The *Upcycle: Beyond Sustainability—Designing for Abundance*, these authors present a compelling argument for the concept of "upcycling" [11]. That is, reusing materials in products that end up and lower on the value chain.

Working through the World Business Council for Sustainable Development, companies like Caterpillar, Coca-Cola, General Motors, Procter & Gamble, and others, have concluded that the "…current global consumption patterns are unsustainable." The Council has called for business to become more proactive in addressing more sustainable methods of consumption through innovation, improved communications, and working in partnerships with governments and other stakeholders [12].

Product stewardship has become the centerpiece of solid waste legislation in many Canadian provinces and the European Union, which has created a process for manufacturers or brand owners to set aside funds to invest in education, research, market development, and collection and processing infrastructure to reduce packaging and take back these materials. In the United States, California, Maine, Minnesota, Oregon, Rhode Island, and Washington have already promulgated product stewardship legislation.

References

[1] US Environmental Protection Agency. Municipal solid waste generation, recycling and disposal in the United States: facts and figures for 2010, <http://www.epa.gov/wastes/nonhaz/municipal/pubs/msw_2010_rev_factsheet.pdf>; 2011. Accessed January 15, 2013.

[2] Conference Board of Canada. How Canada performs, <http://www.conferenceboard.ca/hcp/details/environment/municipal-waste-generation.aspx>; 2013. Accessed March 5, 2013.

[3] US Environmental Protection Agency. Beyond RCRA: waste and materials management in the year 2010. US Environmental Protection Agency; 2002. Washington, D.C.

[4] Los Angeles County Solid Waste Committee/Integrated Waste Management Committee. Task force adopts key definitions and new solid waste paradigm, MSW management, October 31, 2012. Accessed April 3, 2013.

[5] Yergin D. The prize: the epic quest for oil, money and power. New York, NY: Simon and Shuster; 1992.

[6] National Research Council America's climate choices: panel on advancing the science of climate change. Washington, DC: The National Academies Press; 2010. Accessed March 6, 2013.

[7] Cook J. Climate research nearly unanimous on human causes. Environ Res Lett 2013. Accessed March 5, 2013.

[8] Weitz KA, et al. The impact of municipal solid waste management on greenhouse gas emissions in the United States. Air Waste Manage Assoc 2002;52:1000–11. Accessed March 6, 2013.

[9] DSM Environmental Services. Recycling economic information study update: Delaware, Maine, Massachusetts, New York, and Pennsylvania, prepared for the northeast recycling council; 2009.

[10] McDonough Braungart M. Cradle to cradle: remaking the way we make things. New York, NY: North Point Press; 2002.

[11] McDonough Braungart M. The upcycle: beyond sustainability—designing for abundance. New York, NY: North Point Press; 2013.

[12] World Business Council for Sustainable Development. Sustainable consumption facts and trends from a business perspective; 2008.

References

[1] US Environmental Protection Agency. Municipal solid waste generation, recycling, and disposal in the United States: facts and figures for 2010. <http://www.epa.gov/waste/nonhaz/municipal/pubs/msw_2010_rev_factsheet.pdf>; 2011. Accessed January 19, [illegible]

[2] Conference Board of Canada. How Canada performs. <http://www.conferenceboard.ca/hcp/details/environment/municipal-waste-generation.aspx>; 2011. Accessed March 5, [illegible]

[3] US Environmental Protection Agency. Beyond RCRA: waste and materials management in the year 2010. US Environmental Protection Agency, 2003. Washington, DC.

[4] Los Angeles County Solid Waste Commission/Integrated Waste Management Committee. Task force adopts key definitions and new solid waste paradigm, MSW management; October 31, 2012. Accessed April 3, 2013.

[5] Yergin D. The prize: the epic quest for oil, money and power. New York, NY: Simon and Schuster; 1992.

[6] [illegible]

[illegible] in The United States. [illegible] Waste Management [illegible] 2009 [illegible] Accessed March [illegible]

[9] EPA, Environmental Council of States. Recycling economic information study update [illegible]

[11] M [illegible]

[12] World Business Council for Sustainable Development. Sustainable consumption facts and trends from a business perspective; 2008.

2 Policy Implications for Solid Waste and Recycling

City of Santa Monica

The City of Santa Monica, which is located near Los Angeles, has prided itself as an environmental leader having implemented a variety of cutting edge environmental programs over the last few decades. Although the City had reached a reported 75% waste diversion rate by 2009, City political leaders wanted the City to go well beyond these limits. At that time, the City Council directed staff to develop a zero waste strategic plan. Joining forces with its private partners, Southern California Disposal and Allan Company, the City held numerous public meetings to establish a zero waste strategic planning process to identify new programs, City ordinances, product stewardship, and expanded operational requirements that would enable the City to achieve a zero waste goal of 95% diversion by 2030 and a per capita disposal rate of 1.1 pounds per person per day [1].

The following 10 guiding principles (Figure 2.1) were established that would provide a framework for the policies, actions, and actions identified for implementation:

1. *The health of the community and the environment guides all policy decisions*: Santa Monica is committed to protecting, preserving, and restoring the natural environment and safeguarding the health of all members of the community. All programs and policy decisions related to achieving our zero waste goals will be developed based on these commitments.
2. *The hierarchy for managing discarded materials is to reduce, reuse, recycle, and then recover*: The City has adopted an environmental hierarchy for "highest and best use" of discarded materials. It will follow this hierarchy by prioritizing waste prevention and reduction, then encouraging reuse prior to treatment through recycling and composting. The City will recover energy and economic value from residual materials that cannot be recycled or composted through environmentally sound treatment prior to disposal.
3. *Economic and social benefits are integral to zero waste*: The programs and policies in this plan will promote economic benefits, including job creation, cost savings, and business opportunities, and will ensure that inequitable burdens are not placed on any geographic or socioeconomic sector of the population.
4. *The City leads by example*: Santa Monica will model the behavior it seeks from residents, businesses, and institutions by incorporating zero waste principles into local policies and operations; through advocacy for zero waste policies at the regional, state, and federal level; and through support to the community in striving for zero waste.
5. *Brand owners, producers, and manufacturers contribute to the management of their products and packaging*: The City will pursue policies at all levels of government (state/

Solid Waste Recycling and Processing. DOI: http://dx.doi.org/10.1016/B978-1-4557-3192-3.00002-6

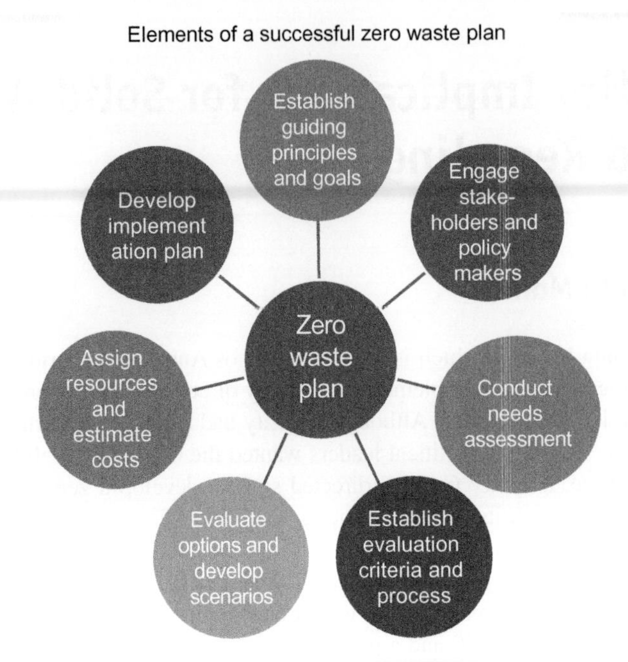

Figure 2.1 Zero waste guiding principles. (Michelle Leonard)

regional/local) for producers to take responsibility for the end-of-life management of products and packaging. Consumers need to be part of the solution and will be educated on alternative purchasing practices.

6. *Regional partnerships leverage the City's efforts in pursuing zero waste*: Santa Monica recognizes that to be successful it must work with other cities in the region to promote a robust zero waste infrastructure and culture. The City will collaborate on programs, public outreach, legislation, infrastructure, and new technologies that help to achieve zero waste goals throughout the region.
7. *Municipal management of local collections and processing programs ensures local control and responsiveness*:—The City's operations are a model of materials management. New programs and infrastructure will be developed to ensure that residents, businesses, and visitors become zero waste champions.
8. *Education, outreach, and marketing are essential to achieving cultural change*: Santa Monica recognizes the power of community-based social marketing, social networks, community organizing, and grassroots support for its education and outreach programs. The City will empower the community by providing tools and techniques for enabling the culture to achieve zero waste.
9. *Research and development of new technologies, collection systems, and infrastructure are needed to maximize diversion of discarded materials*: The City recognizes current approaches to managing materials are inadequate to maximize diversion from disposal. The City will closely monitor new developments and will invest in pilot programs and facilities for achieving zero waste.
10. *Local market development for reusable and recyclable materials ensures sustainability*: Santa Monica is part of the global economy but recognizes that local business

development is the key to a sustainable community. Wherever possible, the City will invest in local reuse and recycling markets and support local businesses in providing markets for discarded materials.

City of San Jose

In October 2007, Mayor Chuck Reed formalized a "Green Vision" for the City, which among other things provides a long term, comprehensive plan for achieving sustainability through improved energy efficiency and a reduction of waste produced by City residents and businesses [2]. In adopting this Green Vision, the City adopted the following 10 overall goals within 15 years:

1. Create 25,000 clean tech jobs as the world center of clean tech innovation.
2. Reduce per capita energy use by 50%.
3. Receive 100% of its electrical power from clean renewable sources.
4. Build or retrofit 50 million square feet of green buildings.
5. Divert 100% of waste from landfills and convert waste to energy.
6. Recycle or beneficially use 100% of its wastewater.
7. Adopt a general plan with measurable standards for sustainable development.
8. Ensure that 100% of public fleet vehicles run on alternative fuels.
9. Plant 100,000 new trees and replace 100% of its streetlights with smart, zero-emission lighting.
10. Create 100 miles of interconnected trails.

Importantly, for our discussion, the City established the following specific objectives for solid waste:

- 75% division by 2013
- Zero waste by 2022

"Zero waste" is defined by the City as a basic change in perspective. That is,

> *... rethinking what we have traditionally regarded as garbage and treating all materials as valued resources instead of items to discard. Zero waste entails shifting consumption patterns, more carefully managing purchases, and maximizing the reuse of materials at the end of their useful life.*

In light of this change in direction, the City established the following zero waste principles:

- Improve "downstream" reuse and recycling of end-of-life products and materials to ensure their highest and best use.
- Pursue "upstream" redesign strategies to reduce the volume and toxicity of discarded products and materials while promoting less wasteful lifestyles.
- Support the reuse of discarded products and materials to stimulate and drive local economic workforce development.
- Preserve land for sustainable development and green industry infrastructure.

Over the past 5 years, the City has started working with its existing solid waste franchise haulers to improve recycling opportunities for both residents and businesses such as expansion of curbside collection of recyclables and organics. In 2010, the City entered into a contract with a private vendor, Zero Waste Energy Development, for construction and operation of a dry anaerobic composting facility to process up to 80,000 tons of food waste, yard trimmings, and other compostable materials and produce a compost product as well as energy from the resulting methane gas.

Metro Vancouver

Metro Vancouver (Metro) is responsible for providing regional services, which include water, wastewater, solid waste, and housing services to some 24 local governments and nearly 2.4 million residents in the Greater Vancouver metropolitan area. This area is the third largest metropolitan area in Canada.

Building a sustainable community has become an overarching goal of Metro Vancouver's solid waste management strategy in its Integrated Solid Waste and Resource Management Plan (ISWRMP). In 2011, Metro Vancouver adopted this new Plan with the following four goals [3]:

- Goal 1: Minimize waste generation.
- Goal 2: Maximize reuse, recycling, and material recovery.
- Goal 3: Recover energy from the waste stream after material recycling.
- Goal 4: Dispose all remaining waste in landfill, after material recycling and energy recovery.

These sustainability principles have provided guidance for the overall waste generation goals: reduction by 10% of 2010 volumes, per capita, by 2020; up to 70% by 2015; and aiming for 90% by 2020. As of 2013, about 55% of Metro's waste is recycled.

To implement the ISWRMP, Metro has strongly advocated the development of the extended producer responsibility mandates being promulgated by the British Columbia Provincial Government (BC) and the enactment of disposal bans on certain waste streams. Since food wastes comprise 21% of the region's waste stream, these materials are being targeted for enhanced diversion through the implementation of a weekly, organics curbside collection, and reducing regular garbage pickup to only once every 2 weeks. Metro is currently investing public dollars for a new organics transfer facility to hold organics until they can be transferred to a private, regional composting facility.

Further, the ISWRMP also mandates the continued operation of a waste-to-energy facility in Burnby and establishment of an additional 500,000 metric tonnes of waste-to-energy capacity with energy recovery in one or more facilities. This latter project is currently in procurement.

New York

Similar to most US state governments, New York, through its Department of Environmental Protection (DEQ), prepared and adopted a comprehensive "resource recovery" plan in 1978, pursuant to Resource Conservation and Recovery Act (RCRA). This Plan has been revised and amended several times over the past 30 years to reflect the environmental consequences of solid waste mismanagement such as unlined landfills, the development of infrastructure to transfer waste for long-haul transport out of state, and the promotion of municipal waste combustion.

Foremost among the events in state solid waste planning during this period was the experience with the Mobro 4000 barge, which set sail from Islip, New York carrying baled municipal solid waste (MSW) for a pilot project in North Carolina. This now infamous garbage barge's month's long saga to find a home to Central America and then back to New York, where it was ultimately incinerated, publicized what was wrong with solid waste planning in New York and the nation. Ultimately, this experience resulted in the adoption of the 1987 Plan, which contained important goals, including a State goal to "reduce, reuse, or recycle 50% of the State's MSW waste stream," and a preferred hierarchy of solid waste management methods:

- First, to reduce the amount of solid waste generated.
- Second, to reuse material for the purpose for which it was originally intended or to recycle the material that cannot be reused.
- Third, to recover, in an environmentally acceptable manner, energy from solid waste that cannot be economically and technically reused or recycled.
- Fourth, to dispose of solid waste that is not being recovered, by land burial or other methods approved by the Department.

Since the adoption of the 1987 Plan, the State's recycling rates have been static or increased only marginally in recent years, although local recycling programs have captured more volumes of recyclable materials. Currently, DEQs figures show that nearly 65% of the total materials managed in New York, and 80% of MSW end up in MWCs and landfills, although landfilling (either in state or out of state) is the solid waste management method statutorily of last resort.

This prompted the DEQ to evaluate development of a new approach in its *Beyond Waste Plan*, to shift from focusing on "end-of-the-pipe" waste management techniques to "looking more comprehensively at how materials that would otherwise become waste can be more sustainably managed through the State's economy [4]." This change in thinking was prompted by several national and regional studies that showed more jobs were created in the "recycling industry" (including collection, processing, and manufacture) than traditional landfill and waste combustion facilities. Further, this change in philosophy by DEQ was central to the State's goal to reduce demand for energy, reduce dependence on disposal facilities, minimize emission of GHGs, and create "green jobs."

New York's new plan is designed to incorporate the following elements to move the State toward what they define as a "sustainable materials economy":

1. *Waste prevention*: Creating and implementing a combination of policies and programs aimed at reducing the volume and toxicity of waste generated and disposed such as:
   - Packaging reduction through stewardship and other means.
   - Extended producer responsibility for key material streams.
   - Purchasing and practices that advance sustainability goals.

- Community outreach and education.
- Incentive for waste prevention through volume-based pricing for waste management programs such as Pay-As-You-Throw (PAYT) or Save Money and Reduce Trash (SMART).

2. *Reuse*: Supporting an expanded infrastructure to redirect items that still have a value for their original intended purpose from those who no longer need them to individuals and entities that can put them to good use.
3. *Comprehensive recycling*: Including more materials and more places; improve education and enforcement to achieve greater participation and greater capture of targeted recyclables in all generating sectors; develop local markets for both traditional recyclables and new materials targeted; and support a manufacturing base that can utilize recycled materials.
4. *Recovery of organics*: Creating a combination of policies and programs to expand backyard composting, onsite composting at institutions and large generators, and develop greater collection and recovery infrastructure for commercial, institutional, and residential food scraps and yard trimmings.
5. *Beneficial use*: Developing policies and programs to redirect items that still have value for uses other than their original intended purpose.
6. *Best residual management strategies*: Advancing policies that ensure adequate capacity of the most environmentally sound and most sustainable means of disposal for the waste that cannot be reduced, reused, composted, or otherwise diverted, placing a preference on disposal methods that recover energy from residual materials.

In summary, *Moving Beyond Waste* develops a series of coordinated programmatic themes and menus of options for New York's communities and solid waste agencies toward a sustainable materials management. In light of the goals noted earlier, the Plan seeks a progressive reduction of MSW destined for disposal to reach the ultimate goal of reducing disposal to 0.6 pounds per person per day by 2030. As the Plan's authors indicate, achievement of this goal will require the involvement and partnership with all "players in the production and supply chain—product manufacturers, distributors, retailers, consumers, and government" and increased investment in recycling and distribution infrastructure.

Oregon

In December 2012, the Oregon Environmental Quality Commission (DEQ) adopted a long-term vision for materials management in the State—2050 Vision and Framework for Action [5]. The resulting document serves as an update to Oregon's State Integrated Resource and Solid Waste Management Plan (1995–2005) to guide state policy. Similar to other state solid waste plans implemented after RCRA, the previous plan focused on "end-of-life" issues such as remediation of poorly sited landfills, landfill closure, and development of state guidelines for landfill siting and operations, and management of the various waste fractions of the typical municipal waste stream.

As noted in the preamble of the State Plan, its objective is to go beyond the focus on managing discards. Rather the plan's authors have utilized the broader definition of materials management to help guide state policy and programs. The Plan envisions Oregon in 2050 where:

- Producers make products sustainable. Every option is a sustainable option.
- People live well and consume sustainably.
- Materials have the most useful life possible before and after discard.

The Plan lists 50 actions that the State will employ in helping shift to a materials management paradigm in *four key areas:*

1. *Foundations*: These include items such as setting goals and measurement guidelines and establishing research programs to evaluate barriers for repair and reuse of materials, and better ways of collecting and separating recovered materials.
2. *Policies and regulations*: These are designed to establish incentives for sustainable consume and corporate patterns. Oregon envisions such policies as credible ecolabels, hazard information for chemicals, directing materials to the highest and best uses, and extended producer responsibility (take-back) programs.
3. *Collaboration and partnerships*: Achieving the goals envisioned by the Plan will require collaboration with all sorts of partners and stakeholders.
4. *Education and information*: Lastly, dissemination of well-researched materials will allow consumers and businesses to embed sustainable policies in day-to-day living or business operations.

Vermont

In 1987, the State of Vermont adopted the State's Waste Management Law (Act 78) to address growing concerns about the mismanagement of "discarded materials." An initial State solid waste plan was adopted in 1989 and readopted with minor modifications in 2001 and 2006. In 2007, a legislative mandate required the Agency for Natural Resources (ANR) to evaluate the effectiveness of the current Plan at that time and develop a new vision for waste management for Vermont based upon its findings.

This report, entitled, *Life Beyond Garbage*, presented ANRs vision for the future of materials management by promoting waste prevention as the "least costly and most beneficial method to protect human health and the environment [6]." That report focused on the following waste reduction strategies of materials:

- Recyclables
- Organics
- Construction and demolition debris (C&D)
- Household hazardous waste and conditionally exempt generator wastes
- Septage and sludges.

Key priorities for Vermont's management of these materials included the following:

- The greatest extent feasible reduction in the amount of waste generated.
- Materials management, which furthers the development of products that will generate less waste.

- The reuse and closed-loop recycling of waste to reduce the greatest extent feasible the volume remaining for processing and disposal.
- The reduction of the State's reliance on waste disposal to the greatest extent feasible.
- The creation of an integrated waste management system that promotes energy conservation, reduces GHGs, and limits adverse environmental impacts.
- Waste processing to reduce the volume or toxicity of the waste stream necessary for disposal.

Each of these priorities has been factored into the development of the Plan. Key features of the State's vision include the following facets:

1. *Focus on sustainable materials management*: Focus on "upstream options " preventing or minimizing waste generation while using resources more effectively. In order to remind Vermonters about the expense of operating and managing the current materials management infrastructure, the State will require the implementation of PAYT programs by all municipalities by July 1, 2015. The Plan's authors suggest that assigning an incremental cost for different volumes of waste disposal provides an economic incentive to recycle.
2. *Shared responsibility*: Efforts to engage all residents, businesses, manufacturers, and institutions rather than placing all responsibility on government and municipalities.
3. *Establish statewide generation goal*: Vermonters should consider reducing waste generation.
4. *More integrated statewide strategies and implementation*: Address the disparities between different areas of the State by collaboratively working with solid waste districts, economic developers, waste haulers, processers, and businesses.
5. *Expand current and create new partnerships*: Working with others to implement the sustainable materials vision.
6. *Continue to meet ongoing responsibilities*: Continued ANR oversight to protect human health and the environment.
7. *Expand criteria when making decisions*: Decisions on waste management currently focus on volume, toxicity, and cost. These should be expanded to include other environmental impacts such as energy use and GHG emissions.
8. *Improved measurement methods*: New measurements of success to include: the number of business start ups, jobs created, businesses incorporating recycled materials into their products, reductions in materials and energy consumed, and reduction in GHGs emitted.

References

[1] City of Santa Monica, Zero Waste Strategic Operations Plan, Prepared by HDR; 2013.
[2] City of San Jose, Zero Waste Strategic Plan, Environmental Services Department, November 2008. San Jose, California.
[3] Metro Vancouver, Integrated Solid Waste and Resource Management for the Greater Vancouver Regional District and Member Municipalities, Metro Vancouver; 2010. Vancouver, Canada.
[4] State of New York, Beyond. Waste: a sustainable material management strategy. Department of Environmental Conservation; 2010. Albany, New York.
[5] Oregon Environmental Quality Commission. Materials management in Oregon—2050 vision and framework for action. State of Oregon: Department of Environmental Quality; 2012. Salem, Oregon.
[6] Vermont Agency of Natural Resources. Materials management in Vermont: history of materials management and planning update. Vermont: Department of Environmental Conservation; 2010. Montpelier, Vermont.

3 Collection Approaches

Waste Reduction and Reuse

The following section provides a brief discussion on source reduction and reuse, including examples of how communities are encouraging residents to rethink what waste is and to aim toward the concept of "zero waste." Source reduction and reuse involves reeducating municipal staff and residents with the goal of optimizing, to the fullest extent possible, the reduction of "waste" materials at the source or the productive reuse of those materials we now consider as waste.

Waste Reduction

Activities and practices that reduce the amount of wastes that are created are usually classified by solid waste professionals as "waste reduction." Waste reduction differs from the other two waste diversion techniques (recycling and composting) because the other methods deal with wastes after the wastes have been generated.

Waste reduction is the highest priority for solid waste management according to the solid waste hierarchy in the United States and is preferred over recycling and composting because the social, environmental, and economic costs are typically lower for waste reduction [1]. All three methods avoid the cost of disposing the diverted materials as garbage, but recycling and composting frequently require significant additional expenses for collecting and processing the materials. Importantly, efforts to reduce and reuse waste translate directly into cost savings as the disposal tonnage and associated costs are reduced. Collection costs can also potentially be reduced.

Source reduction is dependent on several factors including:

- Changing the usage and purchasing habits of residents and the community.
- Changes that businesses undertake voluntarily to reduce the amount of potential waste material associated with products.
- Increasing internal reuse of materials, donations, or exchange of old for new items.

These ideas are discussed further in the following sections.

Extended Producer Responsibility

Briefly, extended producer responsibility (EPR) is a general policy approach which aims to shift the cost of managing consumer packaging from local solid waste agencies to those manufacturers who are producing these products. Those promoting EPR

Solid Waste Recycling and Processing. DOI: http://dx.doi.org/10.1016/B978-1-4557-3192-3.00003-8

assert four major advantages for EPR as a preferred policy approach for end-of-life management for packaging and printed paper [2]:

- EPR causes producers to change packaging design and selection, leading to increased recyclability and/or less packaging use.
- EPR provides additional funds for recycling programs, resulting in higher recycling rates.
- EPR improves recycling program efficiency, leading to less cost, which provides a benefit to society.
- EPR results in a fairer system of waste management in which individual consumers pay the cost of their own consumption, rather than general taxpayers.

In the United States, more than 70 producer responsibility laws have been promulgated in 32 states including 10 categories of consumer products such as automobile batteries, mobile phones, paint, pesticide containers, carpet, electronics, thermostats, and fluorescent lamps [3]. In recent years, there has been a rising tide of states which have passed e-waste EPRs as a consequence of the rapid replacement of these products. Several states have enacted landfill bans which have had an increasing positive impact of product recycling. However, as of this date, no state has enacted an EPR law of programs extending to packaging or printed paper.

As a result of failed voluntary packaging take-back programs in the Europe, public policies were instituted to require manufacturers to be responsible for these materials. In 1994, the EU enacted the Packaging Waste Directive (94/62/EC) requiring its member states to develop regulations on the prevention, reuse, and recycling of packaging waste. These regulations vary from country to country, but most countries mandate that manufacturers pay some or all of the costs of packaging collection and recycling in the form of producer financing, shared costs, tradable credits, or packaging taxes [2]. Many countries in Northern Europe (Austria and German) have decided to develop collection programs for packaging completely separate from solid waste. Shared systems, which split or share municipal authority with manufacturers, are typical of programs existing in Southern Europe [4].

Figure 3.1 shows packaging recycling rates in European countries as of 2009. The data show that 17 countries target of the Packaging Waste Directive (2004/12/EC) is to recycle at least 55% of packaging waste generated, and two countries missed the 2001 target to recycle at least 25%. The highest performing programs in recent years include Denmark (84%), Belgium (79%), The Netherlands (72%), Germany (71%), and Austria (70%) [4].

EPR legislation is in existence in all of Canada's 10 provinces, with four (Ontario, Quebec, Manitoba, and British Columbia) having programs in place. Again, these programs resulted from similar failures of voluntary take-back programs. Ontario and Quebec require manufacturers to pay 50% of the program costs, Manitoba 80%, and British Columbia 100%. In addition, all provinces have enacted deposit systems for beer containers, with eight provinces having similar deposit laws for soft drink containers.

Product Stewardship

Product stewardship is a voluntary initiative aimed at restructuring the way manufacturers design and market products so that they optimize recycling of materials,

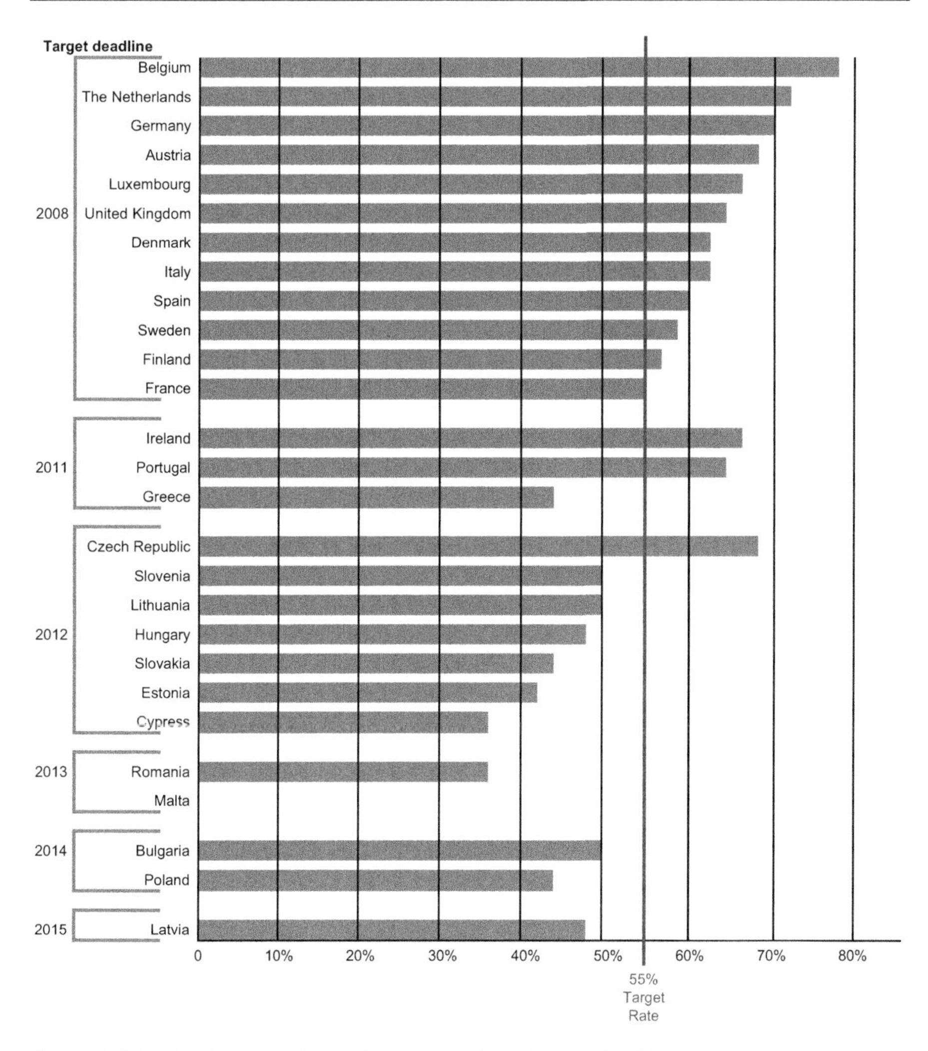

Figure 3.1 Packaging recycling and target rates in EU countries [4].

minimize packaging, and actually design their products in a way that will enable complete recycling of the used product in lieu of disposing the used product. It is essentially a "cradle to cradle" strategy instead of a "cradle to grave" approach.

The Product Stewardship Institute (PSI) is a US, nonprofit membership-based organization, located in Boston. PSI works with state and local government agencies to partner with manufacturers, retailers, environmental groups, federal agencies, and other key stakeholders to reduce the health and environmental impacts of consumer products. PSI takes a unique product stewardship approach to solve waste management problems by encouraging product design changes and mediating stakeholder dialogues. Several states have or are considering initiatives and laws that would encourage or require manufacturers to improve their product designs in this manner.

Economic prosperity has increased per capita spending over the past several years and increased the need for local governments to provide expanded recycling and disposal programs. Product stewardship is a concept designed to alleviate the burden on local governments of end-of-life product management. Product stewardship is a product-centered approach that emphasizes a shared responsibility for reducing the environmental impacts of products. This approach calls on the various waste generators to help minimize their wastes [3]:

- *Manufacturers*: To reduce use of toxic substances, to design for durability, reuse, and recyclability, and to take increasing responsibility for the end-of-life management of products they produce.
- *Retailers*: To use product providers who offer greater environmental performance, to educate consumers on environmentally preferable products, and to enable consumers to return products for recycling.
- *Consumers*: To make responsible buying choices that consider environmental impacts, to purchase and use products efficiently, and to recycle the products they no longer need.
- *Government*: To launch cooperative efforts with industry, to use market leverage through purchasing programs for development of products with stronger environmental attributes, and to develop product stewardship legislation for selected products.

The principles of product stewardship recommend that the role of government is to provide leadership in promoting the practices of product stewardship through procurement and market development. Environmentally Preferable Purchasing (EPP) is a practice that can be used to fulfill this role. EPP involves purchasing products or services that have reduced negative effects on human health and the environment when compared with competing products or services that serve the same purpose. They include products that have recycled content, reduce waste, use less energy, are less toxic, and are more durable.

Table 3.1 provides details on a few examples of programs undertaken by North American producers of consumer packaged goods.

Local, state, and federal government agencies can and do use their tremendous purchasing power to influence the products that manufacturers bring to the marketplace. In the last decade or so, most efforts have focused on encouraging procurement of products made from recycled content. The goal of these procurement programs is to create viable, long-term markets for recovered materials. The EPA has developed a list of designated products and associated recycled content recommendations for federal agencies to use when making purchases. These are known as Comprehensive Procurement Guidelines [5].

To date, EPA has developed more than 60 guidelines that fall into the general categories of construction products, landscaping products, nonpaper office products, paper and paper products, park and recreation products, transportation products, vehicular products, and miscellaneous products [5]. For example, federal agencies are instructed to buy printing or writing paper that contains at least 30% postconsumer recycled content.

Table 3.1 Illustrative Voluntary Product Stewardship Programs

Company or Organization	Program
Coca-Cola Company	Municipal grants for beverage collection bins. Recycling education
Pepsi Cola Company	Dream Machine Initiative to collect beverage containers at away from home locations (bins and kiosks) Reduce weight of Aquafina brand plastic bottle
Target Stores	Store recycling bins for cans, glass containers, plastic bottles, and plastic bags
British Columbia Dairy Council	Deposit return locations for dairy containers
Publix Grocery	Store recycling bins for cans, glass containers, plastic bottles and plastic bags
Best Buy and Staples	Store recycling bins for toner cartridges, e-waste, and cell phones
Walmart	Pressed suppliers like Proctor & Gamble to reduce packaging
Proctor and Gamble	Increased concentration of Tide laundry detergent to sell in smaller packages
Zephyhills Water	Redesigned water bottles for reduced size of cap and plastic in bottle

Zero Waste Initiatives

Many municipalities have investigated and taken on the concept of "Zero Waste." This is currently the most comprehensive all-around way of looking at the concept of source reduction or waste reduction, and there are many sources of information and examples of how a solid waste agency could consider adopting a goal of this type, for advancing waste reduction. It is important to note that "Zero Waste" does not mean that all waste materials will disappear, but that, to the maximum extent possible, source reduction, recycling, and waste diversion will have removed all materials that can be utilized in some way. Instead of seeing used materials as garbage in need of disposal, discards are seen as potentially valuable resources. Zero Waste is a "whole system" approach to resource management that maximizes recycling, minimizes waste, reduces consumption, and ensures that products are reused.

Reuse Programs

Swap shops and/or thrift stores provide a good venue for promoting the reuse of household items. Many communities have informal reuse centers located at their waste collection/drop-off centers, some of which are operated by volunteers. Promoting the reuse of building materials is also prevalent in communities looking for ways to divert materials from disposal. Another reuse avenue becoming more popular is the use of web site exchanges, such as the FreeCycle Network and Craigslist.

Figure 3.2 Charlotte County's C.A.R.E. store. (Roger Lescrynski)

Community Cooperatives and Exchanges

Many communities are initiating cooperatives or exchanges for specific products or interests—such as bicycles or books—in order to facilitate knowledge about a product or subject, assist in repairs, and generally promoting a sense of sustainability. An example of community cooperatives and exchanges is the SWAP Shop operated by the Center for Abuse and Rape Emergencies (C.A.R.E.) in Charlotte County, Florida (Figure 3.2). Project ReUse is a community effort to improve the quality of life in Charlotte County, Florida. C.A.R.E. Project ReUse is a collaborative project with the Charlotte County Environmental and Extension Services to keep usable items out of the local landfill, and at the same time, augment funding to assist victims of domestic violence, sexual assault, and other violent crimes. There are two ReUse stores in the county both colocated with the County's transfer stations used for collecting recyclables. Participants can drop off usable items, such as clothes, furniture, or kitchen supplies free of charge at the C.A.R.E. stores.

Solid Waste Collection Systems

The following section briefly describes general designs of basic solid waste collection.

Self-Haul

Self-haul waste collection requires that the residents or business owners deliver their garbage or recyclables to a central collection location. This option works well for smaller or rural communities who do not have the economies of scale to warrant curbside collection (Figure 3.3).

Figure 3.3 Community recyclables drop-off facility, Island County, WA.

Residential Curbside Collection

Manual

Manual pickup has traditionally been the dominating type of collection since curbside garbage collection services began. This entails garbage collectors jumping off the garbage truck and manually lifting garbage cans to empty. Although this practice has dominated garbage collection for a long time, many communities are moving toward semi-automated and automated systems in order to minimize work-related injuries and for other reasons that are addressed below.

Semi-Automated

Semi-automated garbage collection is a hybrid between manual and automated systems. With semi-automated vehicles, crews wheel the carts to the collection vehicle and line them up with "flippers" (i.e., hydraulic lifting devices mounted on the truck body), activate the lifting mechanism, then return empty containers to the collection point. The use of semi-automated vehicles decreases demand for manual lifting, but it does not eliminate the need for manual labor. However, semi-automated trucks can be significantly smaller and thus are more conducive for neighborhoods with narrow streets and alley ways (Figure 3.4).

Fully Automated

Automated side-loader trucks were first implemented in the City of Phoenix in the 1970s with the aim of ending the back-breaking nature of residential, solid waste collection, and to minimize worker injuries. Since then, thousands of public agencies and private haulers have moved from the once, traditional read-loader method of waste collection to one that also provides the customer with a variety of choices

Figure 3.4 Semi-automated collection vehicle. (SCS Engineers)

in standardized, rollout carts. This has enabled communities throughout the country to significantly reduce worker compensation claims, minimize insurance expenses, while at the same time offering opportunities to workers who are not selected for their work assignment based solely on physical skills.

Modern Application of Automation

For this type of collection system, residents are provided a standardized container into which they place their waste. Residents must place their cart at the curb on collection day. During collection, the driver positions the collection vehicle beside the cart. Using controls inside the cab of the vehicle, the driver maneuvers a side-mounted arm to pick up the container and dump its contents into the hopper of the vehicle. The driver then uses the arm to place the container back onto the curb. Under this type of collection system, the driver is able to service the entire route; the need for additional manual labor is eliminated. The savings in personnel and worker's compensation costs, as well as the increase in crew productivity for automated collection, are well documented throughout the solid waste industry.

Currently, the Waste Equipment Technology Association (WASTEC) estimates that there are roughly about 120,000 solid waste vehicles on the road in the United States with about 25% of all new waste collection vehicles purchased in 2003 (the most recent statistics available) were automated [6]. There is a real sense in the solid waste industry today that automated trucks are significantly increasing their share of the new sales in recent years. Current estimates in the industry suggest that the percentage is more like 50% at the time of this writing. This trend is rapidly increasing as many agencies and private haulers attempt to minimize their increasing insurance costs and more effectively control their cost of labor, while at the same time provide increased customer service levels and opportunities for an aging workforce (Figure 3.5).

Figure 3.5 Example of automated side-loader by the city of Santa Monica, California. (Michelle Leonard)

Advantages of Automated Collection Systems

Some of the general advantages of automated collection often touted by its proponents include the following [7]:

For residents

- Convenient and easy method for residents to dispose of trash.
- Wheeled containers are easier, more maneuverable, and safer for residents because there is no carrying or lifting of heavy trash cans.
- The capacity of most cans provided in these programs is equal to three or four regular trash cans.
- The containers keep rodents and pets out of trash given the tight lids.
- Cleaner, healthier neighborhoods with no litter on streets after pickup.

For the municipality

- Improved collection efficiency and reduced costs.
- Reduced employee injuries.
- Lower turnover rate and increased productivity due to less time missed by injured employees.
- Reduced worker's compensation claims and insurance premiums.
- Reduced rodent problems.

The following sections briefly discuss the general disadvantages of automated collection

Disadvantages of Automated Collection Programs

The primary disadvantage of automated collection is the initial costs of purchasing specialized vehicles and providing carts to homeowners. On an average, the capital cost of an automated side-loader is 20% more than that of a manual rear loader. Additionally, the useful life of an automated vehicle is often less than a rear loader.

Cart costs generally average between $35 and $50 in the United States, each depending on container size. Additional general disadvantages include the following:

1. Automated vehicles require more maintenance than traditional rear end load vehicles and require specialized training of technicians.
2. Homeowners must be educated on where to place bins and what kinds of trash can be collected. Bulky items that do not fit in the cart usually require a separate collection. Overloaded containers, or waste left on the ground, can impact the productivity of collection. Ordinances prohibiting waste left on the ground should be developed, while additional containers can help discourage the practice.
3. Some cities have chosen to automate yard waste collection as part of a transition to automation; however, the size and volume of yard waste makes it less conducive to cart programs and typically requires separate collection with different vehicle types (claw-type trucks or rear end load units). In order to effectively automate yard waste collection, yard waste size limits must be enforced, and alternate methods developed to collect larger, bulk debris items. Some jurisdictions have instituted a volume-based fee for yard waste that exceeds a predefined limit, making the system conducive to automation.
4. Automated collection also does not work in densely populated areas with on-street parking on collection days. However, on-street parking does not prevent a cart-based approach to collection. A hybrid system can be employed in these cases where carts are collected in a semi-automated fashion and many cart-system benefits can still be enjoyed.

Recycling Collection Programs

There are three major steps involved in implementing a recycling program: collecting the recyclables, processing the material, and then getting the material to market. The practices involving the first step are explained in more detail below. The marketing aspect is further discussed in Chapter 5.

Recyclables can be self-hauled, meaning residents and business owners collect and carry their recyclables to a designated drop-off area, or they can be collected curbside, with either an automated truck or manually by solid waste collection workers. The following section describes the different types of collection and processing options available for recycling.

Self-Haul

Many communities choose to have recyclables self-hauled by their residents and business owners to a central drop-off location, such as a transfer station or to several recycling drop boxes located around town (Figure 3.6). This option works well for smaller or rural communities who do not have the economies of scale to warrant curbside recyclable pickup. However, if these drop-off areas are not manned, there is a high likelihood of contamination with typical municipal solid wastes (MSWs).

Drop boxes are designated areas in which residents or businesses can deposit their solid waste and/or recyclables. The drop boxes can accept single materials or

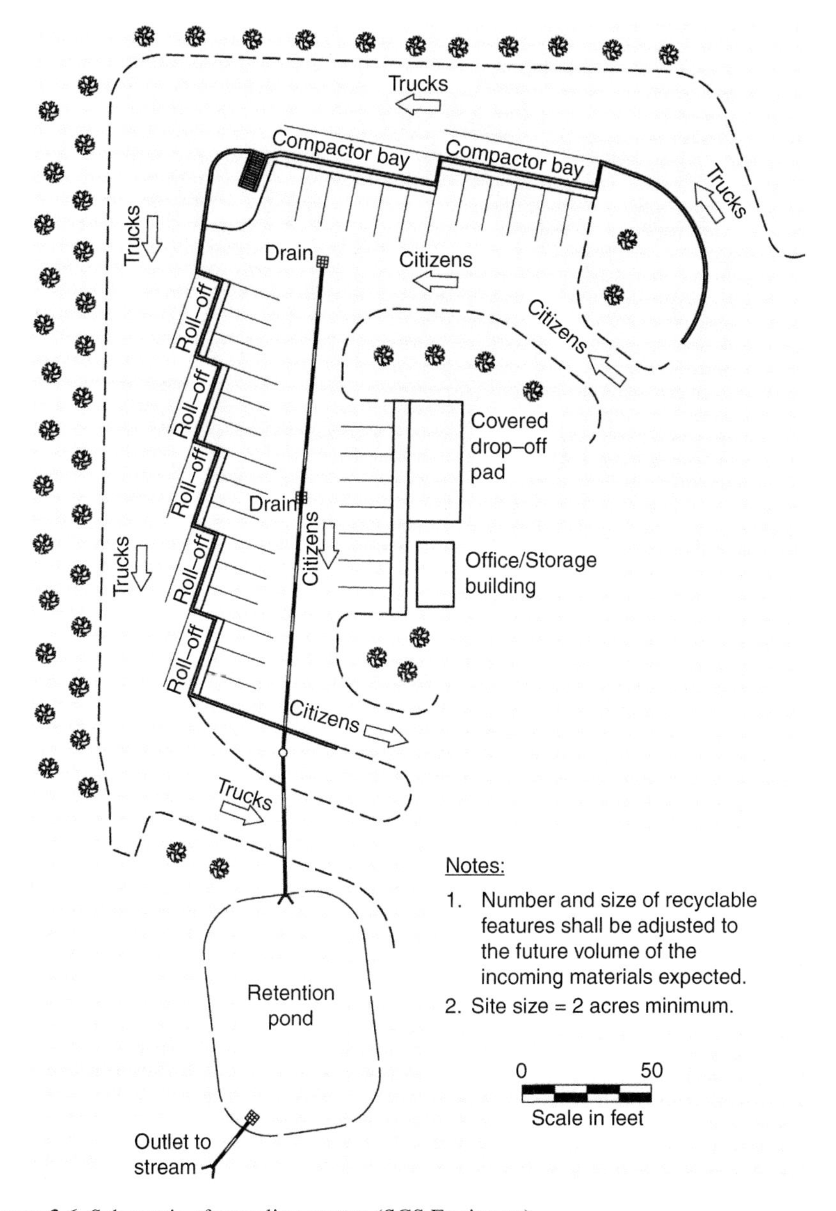

Figure 3.6 Schematic of recycling center. (SCS Engineers)

multimaterials. The drop box may vary from a trailer with designated holes for different material types (Figure 3.7), to a large compactor in which waste is deposited. In many instances, drop boxes are not manned and thus there is a high likelihood that contamination can occur.

Figure 3.7 Community drop box.

Community Transfer Station

Transfer stations are utilized by both self-haul customers and private haulers who bring waste for disposal. In most cases, solid wastes are unloaded into transfer trailers for transport and disposal off-site. Most transfer stations accept various material types from regular MSW to household hazardous waste and recyclables. In most cases, transfer stations also contain scales for weighing waste. Figure 3.8 is an example of a community transfer station, which is designed with several containers for placing recyclables. In this particular application, residents are allowed to drop off their refuse for free provided they deliver recyclables to the transfer station.

Curbside Recycling

Generally, curbside recyclables collection programs are classified into the following two categories: curbside sort and single-stream recycling. These are briefly discussed in the following sections.

Curbside Sort

Under the "curbside sort" method, recyclables deposited curbside in small recycling bins ("blue boxes"), normally 14–25 gallon in size, are manually source-separated by collectors into the various recycling commodities into a specially designed, recycling collection vehicle. These vehicles have compartmentalized bodies to store the sorted recyclables. Once the vehicle is full, the load of separated recyclables is then transported to an intermediate processing facility for further processing (baling) before transport to market.

A variation of curbside sort is dual-stream recycling collection, which is a recycling scheme in which fiber products (newspaper, cardboard, magazines, office

Figure 3.8 Example of citizen's drop-off center at transfer station. (SCS Engineers)

paper, etc.) are collected in one container by the customer, while containers (metals, plastics, etc.) are placed in another one.

Recyclables that are collected in this way tend to receive a higher per ton revenue due to a cleaner product as opposed to single-stream collection, where contamination is more prevalent. However, dual-stream tends to be more labor-intensive and thus there is a lower recycling capture rate, which results in a higher overall disposal cost. Many older dual-stream, materials recovery facilities (MRFs) are converting to single stream as better separation equipment becomes more economical.

Single-Stream Recycling

Programs that provide residents and business owner's convenience and ease of use tend to achieve higher recovery rates and operational costs. The single-stream collection approach was designed to eliminate the need for customers and/or collection crews to sort recyclables curbside. Communities that have adopted single-stream collection programs have achieved significant increases in participation and tonnage recycled. However, single-stream collection requires single-stream MRF processing, which requires additional equipment and capital. Table 3.2 [8] summarizes some of the advantages and disadvantages of single-stream recycling collection programs.

As with most solid waste management approaches, each community is different and oftentimes a "one size fits all" solution will not work well with implementation of single-stream recycling collection [9]. Implementation for a specific community requires good statistics on the current recycling collection program, the types of new equipment suggested, their capital and operating costs, and the projected ease of transition. The anticipated time to affect the writing and evaluation of a new invitation to bid must be factored into this equation. A good game plan is necessary to help minimize the number of errors and vendor issues [9]. Lastly, a proactive public education program is essential to inform the public how the program will be rolled out, when they will receive new equipment, and when their new pickup schedule will occur.

Table 3.2 Advantages and Disadvantages of Single-Stream Recycling Collection Programs

Advantages	Disadvantages
• More customer convenience since it eliminates the need to sort materials at the point of generation. • Increased participation as a result of greater convenience. • Greater amounts of recyclables collected, although other program changes also contribute to this. • No specialized (i.e., compartmentalized) collection vehicles needed; vehicle payload capacity can be optimized. • No curb sorting allows more homes to be served per route. • Allows for automated collection, which requires smaller labor force and results in fewer worker injuries; lower collection costs. • Potential to add more materials to the program, such as plastics #3–7, all types of fiber, and aseptic packaging. • If wheeled carts are used, reduces scavenging and improves community aesthetics. • Potentially higher diversion rates. • Reduced worker compensation claims. • Potential for adding more material types. • Shorter stops and every-other week collection—less wear and tear and fuel savings. • Fuel savings. • Competitive advantage for providers offering this service (marketable service). • Less litter.	• Higher MRF capital and operating costs (although cost per ton is reportedly lower). • Higher initial capital costs for new carts, different collection vehicles, processing facility equipment, and public education. • Lower per-ton revenue to the local government if quality control is not maintained. • Less quality control at the curb under cart-based system. • Higher percentage of processing residue. • Lower recovery of glass by color. • Potential operational and cost impacts to end users if market specifications are not met. • Processing and overall contract costs may increase. • Potential for less net material recovery. • Competitive disadvantage for small haulers unable to offer single sort. • Once you implement single stream, it is difficult to go back to dual stream. • Potential reduced personal commitment to recycling.

Source: From Ref. [8]

Public Areas and Event Recycling

An area often neglected in formulating recycling programs is that of public area and event recycling. With its considerable public presence, its buildings and facilities, and responsibilities for public space and events, solid waste agencies can take a lead role in promoting recycling and showing that government and public employees make waste diversion a normal part of all activities.

Placing recycling bins prominently alongside trash bins in public areas—such as downtown streets, or in recreational parks, can greatly enhance public area recycling.

Figure 3.9 California Strawberry festival. (SCS Engineers)

These bins should be of a different design from the trash bins to clearly distinguish which materials are accepted. Additionally, providing portable recycling containers free of charge to organizations holding public events throughout the year can also increase recycling during holiday parades, festivals, or for other public events.

As an example, since 2007, the California Strawberry Festival has implemented a comprehensive diversion program for vendors and participants of the festival. The 2-day festival attracts 70,000 attendees, 50 food and beverage concessionaires, and 420 arts and craft booths (Figure 3.9). The program includes the following elements:

- Prepare the plan for the festival's sustainability programs.
- Developed diversion program policies and procedures for staff, vendors, concessionaires, and attendees.
- Identify equipment and products needed to make the event eco-friendly.
- Informed and trained volunteers, vendors, concessionaires, and festival staff on diversion program procedures through presentations and printed materials in contracts.
- Supervise and monitor that volunteers, staff, vendors, and concessionaires are following the diversion procedures.
- Evaluate contamination levels of the recycling containers at the material recovery facility to monitor contamination levels after the event.
- Development of a public awareness campaign to make the recycling program visible through the program, local newspaper, event posters, web site, and Internet social marketing tools (i.e., Facebook).
- Prepare final report on the sustainability program procedures, challenges, recommendations, tons diverted, and diversion results.

E-Waste Collection

Electronic waste or "e-waste" has become one of the most persistent waste problems affecting communities today. Common practices for collecting e-waste include community "roundups" held once or twice a year. Larger communities accept self-hauled e-waste at their transfer station or disposal area year round. Donating used electronics for reuse extends the lives of valuable products. Recycling electronics prevents valuable materials from going into the waste stream. Consumers now have many options to recycle or donate, for reuse, their used electronics. Many computer, TV, and cell phone manufacturers, as well as electronics retailers offer some kind of take-back program or sponsor recycling events.

In New York State, the Electronic Equipment Recycling and Reuse Act (Article 27, Title 26 of the Environmental Conservation Law) was signed into law by the Governor on May 28, 2010. The New York law ensures that every New Yorker will have the opportunity to recycle their electronic waste in an environmentally responsible manner. The New York law requires manufacturers to implement and maintain an acceptance program for the discarded electronic waste. The manufacturers must provide for the convenient collection, handling and recycling or reuse of electronic waste via at least one reasonably convenient method of collection within each county, and within each municipality with a population greater than 10,000 at no cost to the consumer.

In May 2013, New York City launched plans for an electronics recycling for residents of apartments through the city. The program is open to apartment buildings with more than 10 units allowing them to voluntarily sign up to have a collection point or collection day. When the bins are full, they are picked up by a private contractor who has a 15-year contract with the City [10].

Other states in the northeast United States that also have EPR laws in effect include Connecticut, Rhode Island, New Jersey, and Maine. These laws have lead to the formation of collectives which represent groups of manufacturers that provide the collection and/or recycling on behalf of the manufacturers.

In summary, best management practices for e-waste include the following activities:

1. Publication and dissemination of information (including via the Internet) about e-waste, including the donation and reuse options or drop-off and mail in programs.
2. Periodic collection or drop-off at licensed facilities.
3. Availability of additional information that may be needed or requested for making proper disposal decisions.
4. Encouragement for local companies and merchants to provide product recycling or take-back opportunities.

Organic Wastes

The organic fraction of the MSW waste stream, which includes food scraps, yard waste, wood waste, and mixed paper, represents about 40–60% by weight [1]. Multifamily residential units do not generate yard waste and wood packaging, so organic wastes are significantly lower, 15–20% by weight, still not an insignificant amount if the community has a high diversion goal.

According to the US EPA, Americans generated nearly 35 million tons of food waste in 2010, with 97% of it disposed at landfills [1]. As such, many communities in recent years have been evaluating options to handle organics beyond the traditional approach in just supplying information on municipal web sites about the benefits of backyard composting. Further, many state and provincial governments have begun promulgating policies and regulations that target the recycling of organics prompting local solid waste agencies to develop advanced municipal curbside collection programs.

A key question for the local solid waste agency to answer is what types of organic wastes will be targeted for collection and processing [11]. For example, some programs accept food wastes, but do not collect meat or fish wastes due to significant odor and processing issues. For example, the plastic lining in some disposal cups, as well as in coated paperboard products, can pose a contaminant problem for composters. Also, other programs restrict the collection of pet wastes and diapers due to contamination concerns. The following sections briefly discuss some of the facets of these organics recycling initiatives.

Drop-Off Programs

Historically, many rural and smaller communities where residents already self-haul refuse, yard waste drop-off can be the cost-effective way to recover a significant amount of organics. Residents who can conveniently haul their yard clippings and other organic wastes to a nearby drop box will participate at levels similar to curbside collection systems. Also, mobile drop-off centers can help serve a number of adjacent communities, especially if these centers offer reduced or free tipping fees for source-separated organics. Food waste collection at drop-off centers has oftentimes proven a bit more complicated than recycling because the materials cannot sit around as long as stacks of newspapers can, but a convenient network of community locations can overcome the barriers to frequent drop-offs by residents.

Bulk Collection

Another simple collection system for organics is for residents to rake their yard clippings, leaves, and brush into piles on the edge of the curb. Trucks with vacuum equipment can then remove the piles and haul them away. If vacuum equipment is unavailable, the piles must be placed in the street so loaders or sweepers can get access to the piles to remove them. Most local governments have dump trucks and loaders and consider this option a less expensive way to implement a yard waste collection program.

This system would only accommodate yard waste since food wastes handled this way would create too much odor and vector attraction. Piles of yard waste in the street could cause traffic problems as well as plugging municipal storm drains. Wet yard wastes piled in this manner could also produce unpleasant odors.

This method of collection could easily be implemented because it does not require anymore effort on behalf of the participants than what is normally expended taking care of their yards. But, the various negative issues introduced with this method would require careful consideration by decision-makers before implementing.

Curbside Collection Programs

According to a 2013 survey in the United States [12], there are more than 214 source-separated organics collection programs in operation and the effort is gaining traction in recent years. That number is up from only 20 programs in 2005. While each of these programs has its own method for food waste collection, several major trends are apparent in both the residential and multifamily sectors.

Residential

A key challenge to residential collection is assisting residents in getting over the "ick factor" of composting organics. Many misconceptions exist regarding storage of organics in households, including the space requirements, public health risks, odor, and rodent problems. Surveys conducted by several municipalities have noted typical comments such as lack of space, odor problems, and lack of time as the top concerns regarding implementation of a household organics collection program. For example, a 2008 study in King County, WA [13], showed that much of public opinion regarding separation of household organics is based on perception rather than reality, and that the "ick factor" dispels when citizens begin recycling household organics.

Current experience also suggests that a municipality must have a strong outreach effort to educate the general public on household waste management practices and illustrate the link between recycling food scraps and lowering refuse collection costs. Information must be easy to understand and the composting process must be as simple and quick for residents as possible. Innovative outreach efforts include composting workshops, illustrated posters of compostable materials, and images of food waste or recyclables displayed on the sides of collection trucks [13].

Single Family

Currently, single-family residential collection of organics is just in its infancy in the United States. Those communities who are "early adopters" have been faced with a series of implementation decisions such as the type of containers, which will encourage a greater participation, and the frequency of collection. Currently, a major trend in residential collection of organics in the United States is providing a variety of kitchen container to help store organics as a means to assist in the daily collection of food scraps and to increase overall participation, as well as some type of external container that will be for curbside pickup.

Collection of household organics is relatively simple and is performed either by the municipality or a waste collection service subcontracted by the municipality. Household organics are placed in a "third cart" and collected weekly or biweekly at the curb. The organics carts range in size from 18 gallons to 65 gallons, depending on whether the municipality allows commingling of yard waste in the carts. Some municipalities that implement the third cart system are able to realize waste collection cost savings by reducing the amount of refuse collections (moving to biweekly or monthly collections). Table 3.3 provides a comparison of collection containers currently used in the United States for organics.

Table 3.3 Comparison of Interior and Exterior Collection Containers

Container	Advantages	Disadvantages
Bucket (1–5 gallon)	• Easily transportable by the user • Can be placed on the kitchen countertop • Size prevents overloading • Differentiable from refuse container	• Difficult to commingle bulky materials such as plants, cardboard, or paper • Not often used for postconsumer food scraps • Not used as collection container for hauler
Slim Jim	• Size consistent with many liners • Available with rollers • Size prevents overloading • Height allows for food to be scraped off for food prep table • Differentiable from refuse container	• Difficult to commingle bulky materials such as plants, cardboard, or paper • Not used as collection container for hauler • Would require lifting into a collection container
Refuse container	• Size allows commingling of bulky organics • Low cost option • Available with rollers	• Potential confusion with refuse container increases potential for contamination • Greater level of signage and employee communication is needed
Carts (32–64 gallon)	• Can be used for interior and exterior collection • Size allows commingling of bulky organics • Rollers to ease transport to outdoor area • Also good for small generators	• Carts taller than food prep tables • Carts can be too large for use by some generators • Food scraps can exceed weight limits for automated arms or tippers if customers are not properly trained • Larger liners are needed for carts
Dumpsters (1–4 CY)	• Can include bulky organics • Once/week collection feasible • Consistent with typical refuse containers for commercial • Plastic containers available	• Employees may be required to lift material overhead • Some customers may not have space for additional container • Must remove from site for cleaning

Source. From Ref. [14]

A majority of communities that offer organics collection have expanded upon an active yard waste collection program by adding such things as food scrap and soiled paper products. This enables “piggybacking” on existing routes and containers, as well as automated collection vehicles or split body collection trucks. Further,

Table 3.4 Comparison of Liners

Types of Liners	Advantages	Disadvantages
No liner	• Cost effective • Low contamination with conventional plastic bags	• More frequent cleaning of container • Difficult to transfer material into collection container
Plastic bag	• Cleanliness of container • No change in purchasing practices from refuse	• Ergonomic issues • Can increase time to unload materials
Compostable bag	• Cleanliness of container • Minimizes nuisances	• Requires purchasing changes • Higher cost • Can be less sturdy
Kraft paper bag	• Cleanliness of container • Paper bags are easily compostable	• Limits material storage time • Purchasing changes • Higher costs • "Workaround" solution

Source: From Ref. [14]

cocollecting yard wastes and food waste together can help mitigate odor and moisture issues [15].

Lastly, the types of materials collected and ability to accept liner materials to help increase the "cleanliness" of the container (Table 3.4) depend largely in part on the ability of the composting facility to accept and process these materials. For example, meat, bones, and dairy scraps usually attract animals and also tend to generate odors and attract flies. While keeping these materials out of an organics collection program may cut back on odor and pest problems, many communities have found it extremely complicated for residents to keep food discards separate. This has usually required a significant investment in public education.

Multifamily

Traditionally, solid waste professionals group multifamily residences into two different groups [15]:

1. Buildings with four to six residential units.
2. Buildings or apartment complexes with six or more residential units.

Those in the first group can usually be serviced by traditional containers (cans) and vehicles, which are used for single-family residences. For the sake of discussion, high-rise buildings refer to multifamily units that typically rely on traditional roll-offs or compactor units for organics collection. To save space, architects most often use areas within the space of building near load-out areas for placement of these units. Normally, regular refuse is fed by gravity through a chute on each floor that is then conveyed to a wheeled container, which is taken periodically to a central roll-off or compactor in the load-out area of the building.

Apartments with six or less units generally have several multistorey buildings with on-site parking. Each building has a central, screened disposal container where refuse and organics can be dumped. These containers are then normally serviced by a front-loader vehicle.

Communities that have implemented organics collection for high-rise units have provided residents with kitchen containers to collect their food waste separate from their normal housed refuse. This means that the residents must transport their organics to a central collection facility, which is oftentimes has proven inconvenient for the residents.

Surveys of these programs have pointed out the following collection issues [15]:

- Resident inconvenience—For the most part, residents in these high-rise buildings are familiar with trash chutes to dispose of their refuse. Therefore, requiring these individuals to source-separate their food wastes and compostable organics in a separate container and bypass the trash chute in favor of carrying these materials to a basement area would be considered inconvenient to many high-rise dwellers.
- Limited space for the collection container—Having building maintenance to provide a separate collection container for source-separated organics is difficult given the limited storage space.
- Limited space for the building container—Storage space in the high-rise building is oftentimes at a premium. Most often, these areas are designed for a single roll-out container.
- Building custodial space workload—Separate containers for organics collection will require custodians to transport these materials for the central building load-out. These containers will also require regulation sanitation to reduce odors and prevent vectors such as insects and rodents. All of these tasks will require extra manpower needs.
- The "Uck factor" No typical American phrase for garbage that smells pretty bad—Food waste and other organics can result in unpleasant odors especially if plastic liners are needed.
- Costs—The extra labor needs and separate collection service will result in extra sanitation service costs by building management.

Collection Approaches

Historically, there have been three different types of collection approaches used to collect organic wastes from multifamily residences. These are discussed in the following sections.

Source Separation In this approach, residents of these multifamily buildings are given a separate container, usually for their kitchen, to separate and store organic wastes (food waste and compostable paper products). As noted above, the resident uses this container to transport and dispose of these materials into a separate organics building container.

One of the largest organics collection programs for high-rise buildings to date is conducted by the City of Toronto, Canada (see Chapter 9). The City's Green Bin Program allows participants to place organics (fruit and vegetable scraps, paper towels, coffee grounds, diapers, person hygiene products, and pet wastes) in small kitchen bins and then into plastic bags for separate weekly collection along with recyclables. The City has reduced the frequency of waste collection to twice per month. To allow plastic bags, a hydropulper at the City's anaerobic digester is used to separate the plastic bags from the resulting slurry for composting.

The City estimates that its multifamily residents generate an average of 165 pounds per residential unit per year. Building owners are responsible for providing a container size equal to 8 cubic yards for every 1000 units. At the end of 2011, 650 buildings representing 120,000 single units have participated in the City's organics program.

Wet/Dry Under this type of collection approach, residents are requested to separate their solid wastes into two different streams: a "dry stream" consisting of recyclables and other wastes, and a "wet stream," which consists of food wastes, coffee grounds, and food-soiled paper and paper products.

Mixed Waste

This collection approach requires that the residence collects all of their wastes in a single container, typically as they do now. Here, the mixed wastes are delivered to a mixed waste MRF where these materials are separated into recoverable recyclables and organics.

Commercial Generators

Commercial food waste generators can economically profit from diverting their unwanted food to beneficial uses. Many organics collection programs in the United States have focused on these potential generators of organics as the "low hanging fruit" to help ramp up their landfill diversion rates and to begin development of a comprehensive organics recycling program. Potential commercial generators of organic wastes include the following [16]:

- Colleges and universities
- Convention centers
- Farming and agriculture
- Food and beverage product manufacturing
- Grocery stores
- Hospitals
- Hotels
- Office buildings and corporate campuses
- Prisons
- Restaurants
- Schools and school districts
- Sports arenas and stadiums.

While these commercial organics collection programs are still in their infancy in the United States, current experience suggests the following steps to help implement a viable program [17,18]:

- Identify what businesses are generating food discards and target these businesses based on type and size.
- Identify businesses that use food discards (such as composters, vermicomposters, animal feeders, animal feed manufacturers, tallow companies). Finding a composting facility that is permitted to take all types of food will result in greater flexibility and higher diversion. If composting facilities can only take vegetative materials, these materials are still worth targeting.

- Try to make matches and distribute information on users to generators so they can make their own matches.
- Place the highest use value on edible food redistribution. When developing a program, work with and support local food donation organizations to incorporate edible food recovery.
- Work with haulers to develop a collection strategy and financial incentives for participating businesses.
- Put time into working with businesses. Provide monitoring and follow-up. Remind businesses that they reap many benefits from participating, including financial and public relations.
- Conduct outreach and find different ways to promote the program. A brochure can help inform businesses about the program. Health departments and chambers of commerce can help deliver messages to businesses.
- Be flexible. As with any new program, be willing to fine-tune the program to meet the needs of cities and customers. Find out if the level of service is right (such as pickup frequency). If not, make adjustments.
- Use front-end loader trucks to collect food discards. Front-end loader trucks are better equipped to handle heavy containers than rear loader trucks.
- Consider providing biodegradable and compostable bags for customers to line their containers as needed. Bags will keep containers clean and prevent food scraps such as dough from sticking to containers, but they will also add to costs.
- Devote a staff person or employ a consultant to work with generators to set up composting systems at generators' sites.
- Offer seed money to cover part of the cost of equipment for on-site diversion.
- Promote business customer recognition programs via local business associations.

References

[1] US Environmental Protection Agency. Municipal solid waste generation, recycling and disposal in the United States: facts and figures for 2010, <http://www.epa.gov/wastes/nonhaz/municipal/pubs/msw_2010_rev_factsheet.pdf>; 2011. Assessed February 21, 2013.

[2] MacKerron C. Unfinished business: the case for extended producer responsibility for post-consumer packaging, as you sow; 2012.

[3] SAIC. Evaluation of extended producer responsibility for consumer packaging, produced for the grocery manufacturers association; 2012. Washington, D.C.

[4] EUROPEN. Packaging and packaging waste statistics in Europe: 1998–2008, the European organization for packaging and the environment; 2011. Brussels, Belgium.

[5] US Environmental Protection Agency, <http://www.epa.gov/epawaste/conserve/tools/cpg/index.htm>. Accessed January 15, 2013.

[6] INFORM. Greening garbage: trends in alternative fuel use, 2002–2005, New York, New York; 2005.

[7] SCS Engineers. Solid waste collection alternatives assessment study, prepared for the city of Lakeland; 2009. Lakeland, Florida

[8] Kessler Consulting, Inc. Materials recovery facility technology review, prepared for Pinellas county department of solid waste operations; 2009. Clearwater, Florida.

[9] Ross D. Rolling out single stream, resource recycling; 2013, Portland, Oregon. p. 18–26.

[10] Carroll J. New York City plans apartment e-waste collection, waste and recycling news; 2013. April 15.

[11] SWANA Applied Research Foundation. Options for recycling source-separated organics, solid waste association of North America; 2012.
[12] Yepsen R. Residential food waste collection in the US biocycle; 2012.
[13] Belcher S. Overcoming the "Ick factor: increasing participation in food scrap recycling programs in king county, WA, report to the king county solid waste division; 2008.
[14] Pasternak S, Wussow K. Organics recycling for commercial, institutional, and single-family customers, workshop presentation for the north central Texas council of governments; 2011.
[15] SWANA Applied Research Foundation. Collection of organic wastes from high-rise buildings and apartment complexes, solid waste association of North America; 2012.
[16] SAIC. Residential source separated organics tool kit, developed by the Georgia department of community affairs and Georgia recycling coalition; 2012.
[17] Kessler Consulting Inc. Mecklenburg county food waste diversion study, prepared for Mecklenburg county, North Carolina; 2012.
[18] Cal Recycle. Food waste recovery: a model for local government recycling and waste reduction; 2010.

4 Processing Technologies

What Can Be Recovered from Your Waste Stream

A key to technology selection is knowing what portion of your proposed waste stream can be diverted and economically recovered by technology. The means to answering this real question is to conduct a waste composition study in a scientific and statistically accurate manner. The paragraphs below illustrate the use of these techniques for a small community in southeast Alaska, but the methodology employed can be applied to most communities around the world.

Borough of Skagway, Alaska Waste Composition Survey

A survey was conducted in September 2012 to answer the following major questions:

- To estimate types and quantities of recyclable and compostable waste components in the residential waste stream; and
- To identify opportunities for greater waste stream diversion.

The basis for this waste characterization consisted of a sampling event, conducted at the Municipality's incinerator. The data generated by the field activities were used by the Municipality to select appropriate recovery technologies, develop long-term waste management strategies, and to develop benchmarks to evaluate the effectiveness of current recycling programs.

Methods

The methods used to characterize the waste stream generated in Skagway are summarized below. Waste characterization activities were performed by manually sorting samples from residential and commercial solid waste (municipal solid waste, MSW) into distinct waste categories.

Waste sorting was performed at the incinerator during the operating hours of the facility. Given the limited size of the data set, it was important that unrepresentative data were avoided. Public Works Department collection vehicles carrying waste from targeted areas of the Municipality were directed to dump their waste loads near the sorting area. Representatives of the engineering consulting firm manually gathered samples (Figure 4.1) from a random portion of each target load (approximately 200lb per sample) for classification (sorting). Two important procedural factors were considered:

- The target vehicle selected for sampling contained MSW that was representative of the type of waste typically generated in that sector; and
- The process of acquiring the waste sample did not, in itself, alter the apparent MSW composition.

Solid Waste Recycling and Processing. DOI: http://dx.doi.org/10.1016/B978-1-4557-3192-3.00004-X

Figure 4.1 Waste being acquired for sample.

Figure 4.2 Samples being stored from targeted loads.

After being filled with solid waste, the trash cans were weighed and set aside until at least 200 lb from the discharged load had been selected for characterization. This process was repeated until samples had been collected from all of the targeted loads (Figure 4.2).

Figure 4.3 Transport refuse sample to sorting table.

Number of Samples

A total of 12 samples were collected during the week, six from residential routes and six from commercial routes.

Waste Sorting

The sorting and weighing program for samples entailed the use of two employees. During each day of fieldwork, samples were collected from waste loads that were discharged at the incinerator. The basic procedures and objectives for sorting (as described below) were identical for each sample, each day. Sorting was performed as follows:

1. The sort crew transferred the refuse sample onto the sorting table (Figure 4.3) until it was full and then began sorting activities. Large or heavy waste items, such as bags of yard waste, were torn open, examined, and then placed directly into the appropriate waste container for subsequent weighing.
2. Plastic bags of refuse were opened, and sort crew members manually segregated each item of waste according to categories (Figure 4.4) defined in Table 4.1, placing it in the appropriate waste container. These steps were repeated until the entire sample was sorted.
3. At the completion of sorting, the waste containers were moved to the scale where a representative of the engineering consulting firm weighed each category and recorded the net weight on the Sort Data Sheet (Figure 4.5). Measurements were made to the nearest 0.1 lb.
4. After each waste category was recorded, the waste was piled onto the incinerator conveyor belts.

This four-step process was repeated until all of the day's samples were characterized. Waste samples were maintained in as-disposed condition or as close to this state as possible until the actual sorting began. Proper site layout and close supervision of sampling was maintained to avoid the need to repeatedly handle sampled wastes.

Figure 4.4 Waste segregated into categories.

Figure 4.5 Electronic weigh scale.

Table 4.1 Description of Waste Categories

Major Waste Fractions	Waste Component Categories	Examples
Paper	Newspaper/print	Daily, weekly newspapers
	Corrugated cardboard	Packing/shipping boxes
	Magazines/catalogs/ other books	TV guide, periodicals, journals, other paper
	Kraft paper/paperboard	Grocery bags, deli packaging
	High-grade office paper/misc. paper	Copy paper, computer printouts, junk mail, notebook paper
	Wax-coated containers	Milk and juice cartons
Plastic	PET (#1) bottles	Water, soda
	HDPE (#2) bottles	Milk, detergent
	Other (#3 to #7) bottles	Prescriptions
	Jars, jugs, tubs, trays	Yogurt, butter
	Films	Garbage bags, bubble wrap
	Shopping bags	Grocery bags
	Polystyrene	Expanded or regular clamshells, cutlery, cups
	Other rigid plastic	Buckets, storage totes, furniture, toys
Metal	Ferrous cans	Pet food cans, soup cans, aerosols
	Other ferrous	Ferrous scrap metals
	Aluminum cans	Soda cans, beer cans
	Aluminum tin/foil	Tin foil
Organics	Vegetative food	Salads, fruits, vegetables
	Nonvegetative food	Meats, dairy products
	Compostable paper	Tissues, napkins, paper towels
Glass	Glass bottles/jars	Beer, wine
Yard waste	Yard waste	Foliage, lawn clippings, brush/branches
Electronics	Electronics	Cell phones, radios
Paint	Paint	All paint
	Wood/lumber	Forklift pallets
	Furniture	Tables, chairs
C&D and bulky wastes	Concrete/brick/rock/ dirt	Gravel, bricks, stones, broken-up asphalt, concrete
	Sheet rock	Drywall
	Carpet/carpet padding	Carpet and carpet padding
	Shingles	Asphalt shingles
Other MSW	Other MSW	Garbage, misc not characterized above, like clothing, or products that contain combinations of materials, such as frozen juice cans, kitty litter, sweepings, mashed food

Source: From Ref [1].

Members of the sorting crew were fully equipped with high visibility vests, puncture/cut resistant gloves, safety glasses, and tyvek suits. Consistent with good practice in such sampling programs, efforts were made to minimize the sampling bias or other impacts on the integrity of the database. To this end, field sampling had been coordinated to avoid holidays and other out of ordinary events.

Data Reduction

There were 12 samples manually sorted during the September 2012 field activities. Data presented include mean percentages by weight, standard deviations, and statistical confidence intervals (95% confidence interval) for each group of data. Derivation of this data is as follows:

$$\text{Mean}(\bar{X}) = \sum_{i=1}^{n} xi * \frac{1}{n}$$

$$\text{Standard deviation}(s) = \sqrt{\frac{\left(n\sum x^2\right) - \left(\sum x\right)^2}{n(n-1)}}$$

and

$$\text{Upper/lower confidence interval limits} = \bar{X} \pm \left[1.96 * \left(\frac{\sigma}{\sqrt{n}}\right)\right]$$

where

n=number of samples; and
x=sample percentage.

Waste samples were acquired to estimate the Municipality's true residential waste composition (i.e., the proportion of each waste component present in residential waste collected in the Municipality). The mean is the arithmetic average of all data, and the standard deviation is a measure of the dispersion in the data. Together, the mean and standard deviation determine the confidence interval. A 95% confidence interval contains the true proportion of a waste component with 95% confidence interval (i.e., similar studies will produce the same results 95% of the time).

Summary of Results

Residential

Table 4.2 presents a compilation of the six residential waste samples collected. The composition includes confidence intervals based on the number of samples and variability between the samples. Based on the samples collected, the three largest subcomponents, by weight, of the residential waste stream are other MSW (23.4%), compostable paper (8.8%), and vegetative food (8.7%). The three largest recyclable subcomponents are magazines/catalogs/other books (7.1%), office paper/other paper (5.4%), and paperboard (4.5%) (Figure 4.6).

Table 4.2 Residential Waste Composition

Material Components (%)	Mean Composition (%)	Standard Deviation (%)	95% Confidence Limits	
			Lower (%)	*Upper (%)*
Paper				
1 Newspaper/print	2.3	2.4	0.3	4.2
2 Corrugated cardboard	3.0	0.8	2.4	3.7
3 Magzines/catalogs/other books	7.1	4.4	3.6	10.7
4 Kraft paper/paperboard	4.5	0.9	3.8	5.3
5 Office paper/other paper	5.4	2.3	3.6	7.2
6 Wax-coated containers	0.7	0.3	0.5	1.0
Total paper	**23.0**			
Plastic				
7 PET #1 bottles	2.3	1.0	1.5	3.1
8 HDPE #2 bottles	0.7	0.5	0.3	1.1
9 #3 to #7 Plastic bottles	<0.1	<0.1	<0.1	<0.1
10 Jars, jugs, tubs, and trays	1.9	0.6	1.4	2.3
11 Plastic films	7.3	1.8	5.9	8.7
12 Shopping bags	0.6	0.4	0.2	1.0
13 Polystyrene	1.2	0.5	0.8	1.6
14 Other rigid plastic	1.2	1.2	0.2	2.2
Total plastic	**15.2**			
Metal				
15 Ferrous cans	2.2	1.2	1.2	3.1
16 Other ferrous	0.9	2.0	<0.1	2.4
17 Aluminum cans	0.8	0.2	0.7	1.0
18 Aluminum tins/foil	0.4	0.2	0.3	0.6
Recyclable metals	**4.3**			
Organics				
19 Vegetative food	8.6	2.6	6.4	10.7
20 Nonvegetative food	8.7	4.2	5.3	12.1
21 Compostable paper	8.8	1.0	8.0	9.6
Organics	**26.1**			
Glass				
22 Glass bottle/jars	3.5	1.3	2.4	4.6
Yard waste				
23 Yard waste	3.5	2.9	1.2	5.8
Total yard waste	**3.5**			

(Continued)

Table 4.2 (Continued)

Material Components (%)	Mean Composition (%)	Standard Deviation (%)	95% Confidence Limits	
			Lower (%)	*Upper (%)*
Electronics				
24 Electronics	0.4	1.0	<0.1	1.2
Paint				
25 Paint	<0.1	<0.1	<0.1	<0.1
Other MSW				
26 Other MSW	23.4	3.9	20.3	26.5
C&D and bulky wastes				
27 Wood/lumber	0.3	0.6	<0.1	0.8
28 Furniture	<0.1	<0.1	<0.1	<0.1
29 Concrete/brick/rock	<0.1	<0.1	<0.1	<0.1
30 Sheet rock	0.2	0.6	<0.1	0.7
31 Carpet/carpet padding	<0.1	<0.1	<0.1	<0.1
32 Shingles	<0.1	<0.1%	<0.1	<0.1
Total C&D and bulky wastes	**0.6**			
TOTALS	**100.0**			

Source: From Ref [1].

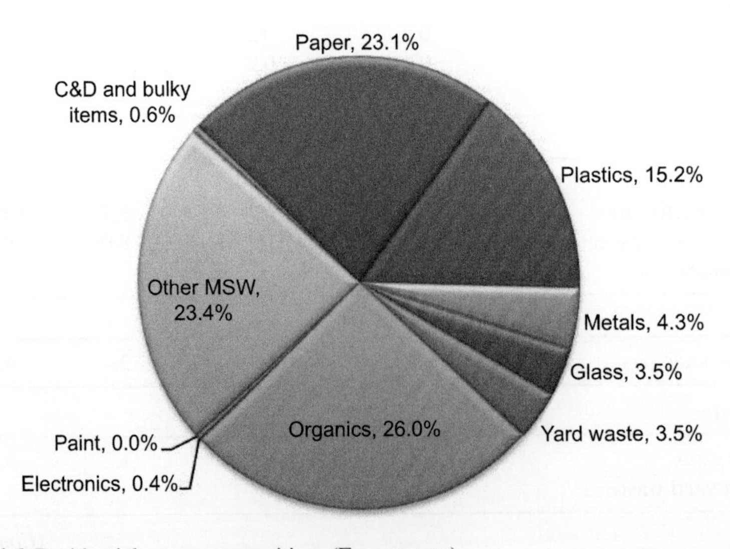

Figure 4.6 Residential waste composition. (From Ref. [1])

Commercial

Table 4.3 presents a compilation of the six waste samples collected. The composition includes confidence intervals based on the number of samples and variability between the samples. Based on the samples collected, the three largest subcomponents, by weight, of the commercial waste stream are other MSW (18.7%), nonvegetative food (13.5%), and vegetative food (10.7%). The three largest recyclable subcomponents are corrugated cardboard (6.3%), glass bottles/jars (5.8%), and paperboard (3.7%).

Table 4.3 Commercial Waste Composition

Material Components		Mean Composition (%)	Standard Deviation (%)	95% Confidence Limits	
				Lower (%)	*Upper (%)*
Paper					
1	Newspaper/print	0.9	0.8	0.3	1.5
2	Corrugated cardboard	6.3	2.9	4.0	8.6
3	Magzines/catalogs/other books	3.0	1.9	1.4	4.5
4	Kraft paper/paperboard	3.7	1.4	2.5	4.8
5	Office paper/other paper	3.4	1.3	2.4	4.5
6	Wax-coated containers	3.6	1.2	2.6	4.5
	Total paper	**20.9**			
Plastic					
7	PET #1 bottles	1.7	0.5	1.3	2.1
8	HDPE #2 bottles	1.4	0.7	0.8	2.0
9	#3 to #7 Plastic bottles	<0.1	<0.1	<0.1	<0.1
10	Jars, jugs, tubs, and trays	1.2	0.7	0.6	1.7
11	Plastic films	8.5	3.2	6.0	11.0
12	Shopping bags	0.5	0.3	0.3	0.8
13	Polystyrene	0.8	0.5	0.4	1.2
14	Other rigid plastic	0.7	0.7	0.2	1.3
	Total plastic	**14.8**			
Metal					
15	Ferrous cans	2.1	1.0	1.3	3.0
16	Other ferrous	0.6	1.0	<0.1	1.5
17	Aluminum cans	1.4	1.1	0.6	2.3
18	Aluminum tins/foil	0.5	0.2	0.2	0.5
	Recyclable metals	**4.6**			
Organics					
19	Vegetative food	8.9	4.5	5.3	12.6
20	Nonvegetative food	13.5	7.0	7.9	19.0
21	Compostable paper	10.7	3.2	8.1	13.2
	Organics	**33.1**			

(*Continued*)

Table 4.3 (Continued)

Material Components	**Mean Composition (%)**	**Standard Deviation (%)**	**95% Confidence Limits**	
			Lower (%)	*Upper (%)*
Glass				
22 Glass bottle/jars	5.8	3.5	3.0	8.6
Yard waste				
23 Yard waste	<0.1	<0.1	<0.1	<0.1
Total yard waste	**0.0**			
Electronics				
24 Electronics	<0.1	<0.1	<0.1	<0.1
Paint				
25 Paint	<0.1	<0.1	<0.1	<0.1
Other MSW				
26 Other MSW	18.7	5.3	14.5	23.0
C&D and bulky wastes				
27 Wood/lumber	1.6	2.7	<0.1	3.8
28 Furniture	<0.1	<0.1	<0.1	<0.1
29 Concrete/brick/rock	<0.1	<0.1	<0.1	<0.1
30 Sheet rock	0.3	0.7	<0.1	0.9
31 Carpet/carpet padding	0.2	0.5	<0.1	0.6
32 Shingles	<0.1	<0.1	<0.1	<0.1
Total C&D and bulky wastes	**2.1**			
Totals	**100.0**			

Source: From Ref. [1]

During field activities it was noted that paper coffee cups (wax-coated containers) and restaurant food waste were very common in the waste stream (Figure 4.7).

Overall Waste Stream

Table 4.4 presents a compilation of the 12 residential and commercial samples collected. This composition is based on the combination of the six residential and six commercial samples. The composition includes confidence intervals based on the number of samples and variability between the samples. Based on the samples collected, the three largest subcomponents, by weight, of the overall waste stream are other MSW (21.1%), nonvegetative food (11.1%), and compostable paper (9.7%). The three largest recyclable subcomponents are magazines/catalogs/books (5.0%),

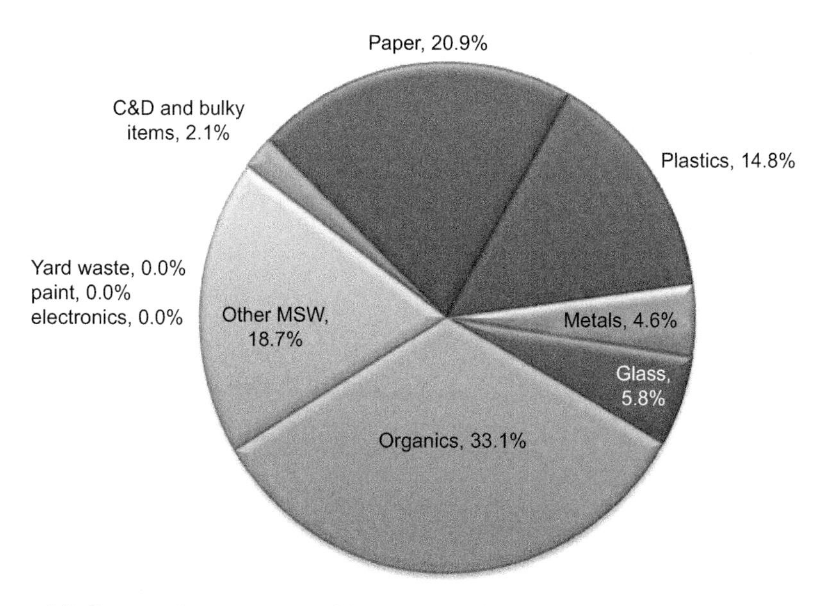

Figure 4.7 Commercial waste composition. (From Ref. [1])

Table 4.4 Overall Waste Composition

Material Components		Mean Composition (%)	Standard Deviation (%)	95% Confidence Limits	
				Lower (%)	*Upper (%)*
Paper					
1	Newspaper/print	1.6	1.8	0.1	3.0
2	Corrugated cardboard	4.7	2.6	2.6	6.8
3	Magzines/catalogs/other books	5.0	3.9	1.9	8.2
4	Kraft paper/paperboard	4.1	1.2	3.1	5.1
5	Office paper/other paper	4.4	2.1	2.8	6.1
6	Wax-coated containers	2.1	1.7	0.8	3.5
	Total paper	**21.9**			
Plastic					
7	PET #1 bottles	2.0	0.8	1.4	2.6
8	HDPE #2 bottles	1.0	0.7	0.5	1.6
9	#3 to #7 Plastic bottles	<0.1	<0.1	<0.1	<0.1
10	Jars, jugs, tubs, and trays	1.5	0.7	1.0	2.1
11	Plastic films	7.9	2.5	5.9	9.9
12	Shopping bags	0.6	0.4	0.3	0.9
13	Polystyrene	1.0	0.5	0.6	1.4
14	Other rigid plastic	1.0	1.0	0.2	1.8
	Total plastic	**15.0**			

(*Continued*)

Table 4.4 (Continued)

Material Components		Mean Composition (%)	Standard Deviation (%)	95% Confidence Limits	
				Lower (%)	*Upper (%)*
Metal					
15	Ferrous cans	2.2	1.1	1.3	3.0
16	Other ferrous	0.8	1.5	<0.1	2.0
17	Aluminum cans	1.1	0.8	0.5	1.8
18	Aluminum tins/foil	0.4	0.2	0.2	0.6
	Recyclable metals	**4.5**			
Organics					
19	Vegetative food	8.7	3.5	5.9	11.6
20	Nonvegetative food	11.1	6.0	6.2	15.9
21	Compostable paper	9.7	2.5	7.8	11.7
	Organics	29.5			
Glass					
22	Glass bottle/jars	4.6	2.8	2.4	6.9
Yard waste					
23	Yard waste	1.8	2.7	<0.1	3.9
	Total yard waste	**1.8**			
Electronics					
24	Electronics	0.2	0.7	<0.1	0.8
Paint					
25	Paint	<0.1	<0.1	<0.1	<0.1
Other MSW					
26	Other MSW	21.1	5.1	17.0	25.1
C&D and bulky wastes					
27	Wood/lumber	1.0	2.0	<0.1	2.6
28	Furniture	<0.1	<0.1	<0.1	<0.1
29	Concrete/brick/rock	<0.1	<0.1	<0.1	<0.1
30	Sheet rock	0.3	0.6	<0.1	0.8
31	Carpet/carpet padding	0.1	0.4	<0.1	0.4
32	Shingles	<0.1	<0.1	<0.1	<0.1
	Total C&D and bulky wastes	**1.4**			
TOTALS		**100.0**			

Source: From Ref [1].

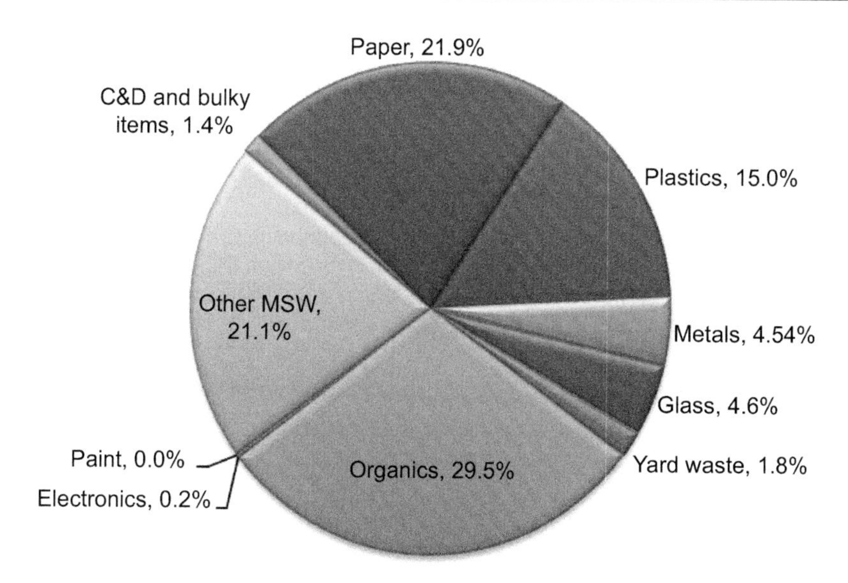

Figure 4.8 Overall waste composition. (From Ref. [1])

corrugated cardboard (4.7%), and office paper/other paper (4.4%). Moisture affects the weights of paper and absorbent materials more than other. Moisture was a factor during the waste composition due to recent precipitation events. Some waste composition studies make adjustments for moisture content to compensate for liquids absorbed by waste materials. Laboratory methods for estimating moisture content are available but are usually expensive and may overestimate moisture. In addition, materials received at disposal facilities are generally measured on an "as is" basis and thus reflect comparable weights as were acquired for this study. Therefore, the designers did not include an analysis of or adjustments for moisture content as part of this study (Figure 4.8).

Diversion Opportunities

A significant portion of the waste stream is compostable or recyclable. Some materials, such as wax-coated paper, other glass, and plastic film (largely plastic bags and packaging), are considered trash since these materials do not currently have obvious markets for recycling or composting. Table 4.5 details the materials included in the compostable, recyclable, and trash classifications used for this section.

The largest diversion opportunities (by weight) for the Municipality are capturing recyclable paper and composting organics. Figures 4.9 and 4.10 portray the waste composition by recyclable and compostable materials. According to the waste

Table 4.5 Compostable, Recyclable, and Trash Classifications for Waste Materials

Compostable	Recyclable		Trash
Compostable paper	Newspaper	Shopping bags	Wax-coated containers
Vegetative food	Corrugated cardboard	Steel food cans	#3 to #7 Plastic bottles
Nonvegetative food	Paperboard/kraft paper	Other ferrous	Plastic films
Yard waste	Office/mixed paper	Aluminum cans	Rigid plastics
Wood/lumber	Magazines/books	Other aluminum	Paint
	PET #1 bottles	Glass bottles/jars	Other MSW
	HDPE #2 bottles	Electronics	Furniture
	Jars/tubs/trays		Concrete/brick/rock
			Sheet rock
			Carpet/carpet padding
			Shingles

Source: From Ref. [1].

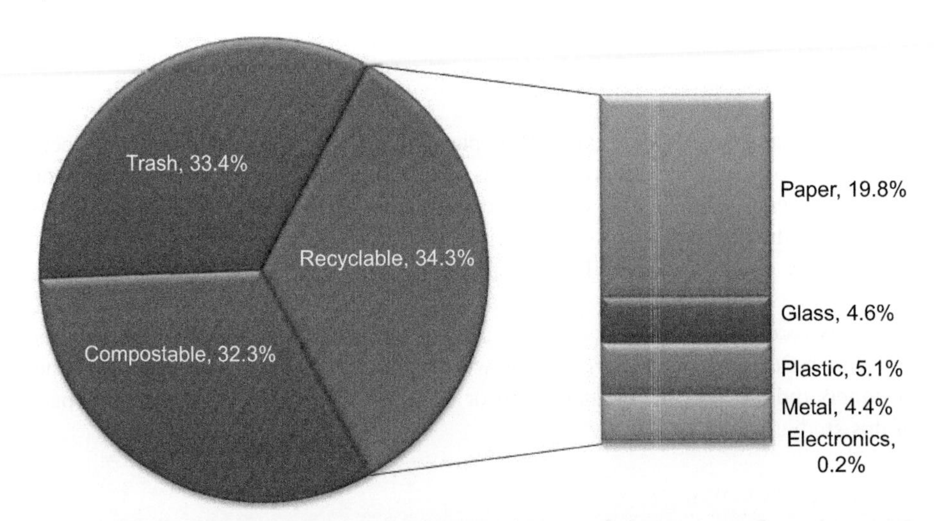

Figure 4.9 Recyclable diversion opportunities—overall waste stream. (From Ref. [1])

characterization, approximately 67% of the overall waste stream is considered recyclable or compostable. Compostable materials such as food waste were more prevalent in the commercial waste stream, and some recyclable materials such as paper were more prevalent in the residential waste stream. The following exhibits are based on the Municipality's overall waste stream (residential and commercial combined).

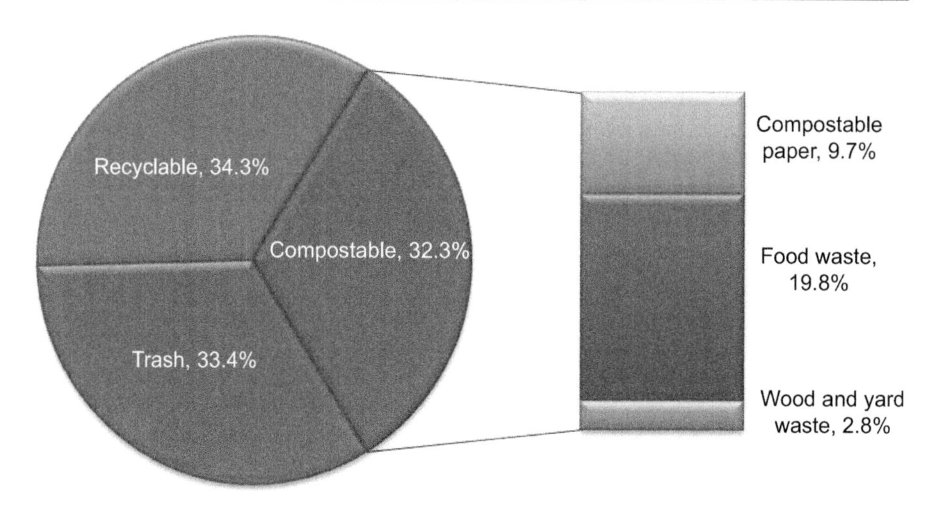

Figure 4.10 Composting diversion opportunities—overall waste stream. (From Ref. [1])

Conventional Processing Technologies

Once recyclables are collected, the materials have to be processed. There are several levels of processing, typically ranging from a small shed with a few balers to a full-scale materials recovery facility (MRF) that separates out the recyclables. A summary of these different technologies is detailed in the paragraphs below.

Source-Separated Recycling Drop-Off Facility

A source-separated recycling drop-off facility provides minimum processing of recyclables. Most of these facilities contain various forms of balers that can condense and bale recyclables by material type. For example, there may be a paper baler and a metals baler onsite. Material delivered here is usually sorted by the person dropping off the recyclables; although some facilities employ workers to assist with this task and still others depend on volunteers. There is very little garbage or contaminants associated with this type of collection system, so they tend to be more acceptable to communities when they are going through the facility siting process.

Some communities elect to have a nonprofit community recycling facility. The general setup of these facilities is to have customers self-haul and source separate materials onsite. Oftentimes, the facility is operated by part-time employees or on a volunteer basis. Funding for the facility is often typically solely dependent upon memberships and/or the sale of recycled material.

Materials Recovery Facility

An MRF is a facility where solid waste or recyclables are sorted to recover materials for marketing to end users such as manufacturers or sent to other recycling facilities for further processing. These facilities have been a product of continuing design improvements over the last two decades and an increasing enhancement of management techniques [2]. Commonly, these facilities are classified on how the materials are sorted and processed:

- Dual-stream MRF
- Single-stream MRF
- Mixed waste MRF.

Current State of MRFs

As shown in Figure 4.11, the growth of MRFs in the United States has risen substantially over the past 20 years from 40 in 1991 to an estimated 736 in 2012. Nearly 60% of these are located in 11 states with the largest populations (California, Pennsylvania, New York, Florida, Minnesota, Texas, Wisconsin, New Jersey, Ohio, Illinois, and Virginia), and almost 70% of these are owned by private service providers, most commonly the largest United States waste haulers (Waste Management, Republic, Progressive).

Newer MRFs are becoming larger to take advantage of economies of scale in operating costs. In the early 2000s, there were only a handful of MRFs processing in excess of 100,000 tons of recyclables per year. However, in recent years, this is no longer the exception and is quickly becoming the state-of-the-art size for newly constructed facilities. Further, the industry has seen consolidation with many of the larger firms merging or buying smaller operations to gain regional market share [2].

Changes in the MSW stream have had an impact on the overall design of MRFs in the United States over the past two decades. The greatest impacts have been seen in the reduction of newspapers and the increase in cardboard in the MSW stream in the United States.

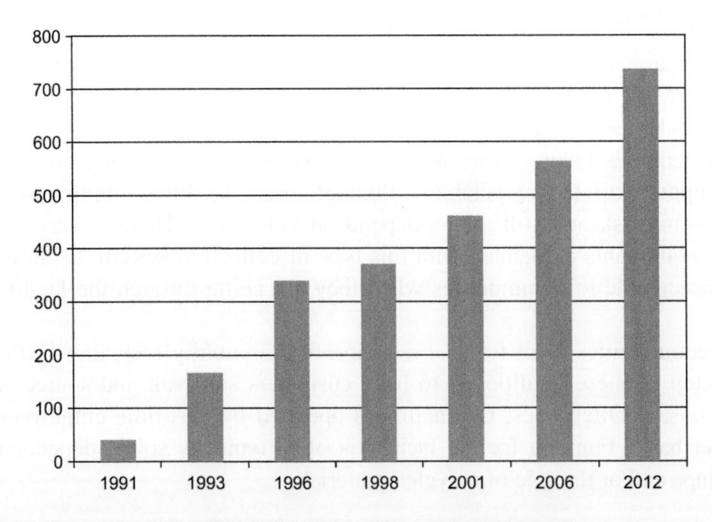

Figure 4.11 Growth in U.S. MRF, 2001–2012. (From Eileen Berenyi)

The loss of advertising revenues in the United States since 2007 due to the Great Recession has had a noticeable impact on the print media world. Newspaper has traditionally been a significant component of the recyclables generated by residences, oftentimes representing 60% of the recyclables by weight. Several major forecasters have predicted the decline in newspaper consumption over the next several decades, suggesting that we are seeing the end of the age of print media [3]. Figure 4.12 graphically portrays the reduction in the magazine and newspaper business.

Figure 4.13 illustrates the increasing percentage of online sales as a percentage of all retail sales. With respect to MRF design, this trend represents itself in the

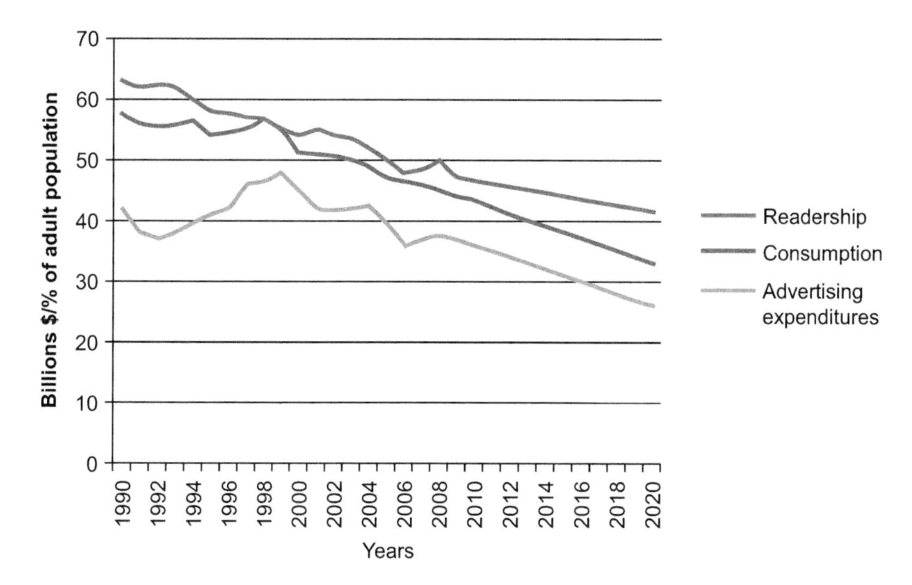

Figure 4.12 U.S. daily newsprint consumption, readership, and advertising expenditures. (From Ref. [4])

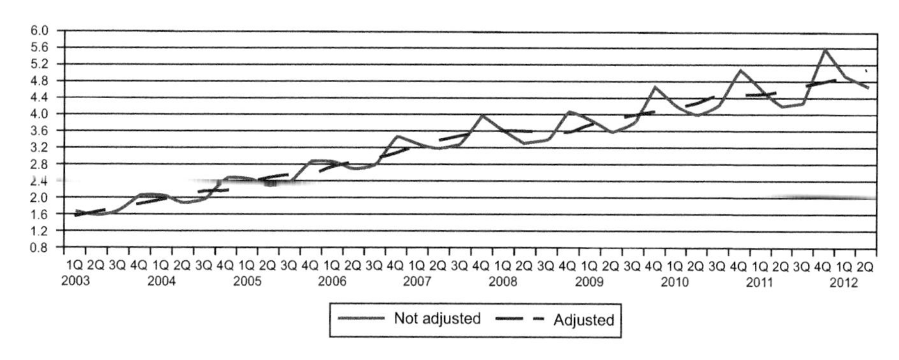

Figure 4.13 Estimated retail e-commerce sales as a percent of total retail sales. (From Ref. [5])

increasing percentage of old corrugated containers (OCC) in the MSW stream as more products are shipped to residences in the United States in OCC.

The MRF market in the United States is also experiencing a significant number of facility modifications to accommodate municipal preferences for single-stream recycling. Many facilities have seen additions of single stream lines or wholesale reconstruction to single-stream processing only. Most industry observers suggest that approximately half of the larger United States MRFs process single-stream recyclables [6]. Baling, as well as loose transfer of single-stream recyclables, for long-haul transport is widely popular in many regions to take advantage of lower operating costs of these larger "super MRFs."

Nonetheless, the debate continues between proponents of single-stream recycling and those who believe that it results in increased residue rates and reduced material quality. Many believe, that while recovered paper from single-stream recycling contain higher levels of contamination than from dual-stream programs, this is caused in part to the decline in the amount of olds newsprint and plastics in the single-stream programs. Industry observers suggest that the issue of contamination in single-stream MRFs will be mitigated by the improvement of separation technologies, the reduction of glass in the incoming waste stream, and extra public education efforts [6]. Lastly, specialized industrial engineering tools such as data logging and financial controls are being included in these systems along with improved indoor working conditions.

Standard Processing Configurations

The following paragraphs briefly describe the standard types of equipment used in most operating MRFs.

Tipping Floor

Materials for processing are delivered to the MRF and offloaded on a paved tipping floor where they are stored inside a covered area to minimize leachate runoff and to keep materials dry. Water can significantly reduce the marketability of the recyclables, particularly paper and cardboard. MRF designers commonly size the tipping floor to accommodate 2- or 3-day supply to enable an adequate supply of materials to either operate during nonscheduled equipment downtime, to accommodate the processing needs of a second shift, and as a buffer for holiday periods.

Not unlike other waste management facilities, MRFs utilize spotters and other laborers to help minimize offloading of potential contaminants or to remove oversize items such as cardboard. The tipping areas are also typically designed with concrete push walls to help protect the building and to help facilitate handling and storage (Figure 4.14).

In-Feed Conveyors

In many MRFs, a front-end loader is used to supply horizontal in-feed conveyors, which are placed below the tipping floor. These conveyors are connected to an incline conveyor running at a slightly faster speed to help spread out the material and

Figure 4.14 Tipping floor, Outagamie County single-stream MRF. (From Phillip Stecker)

deliver the incoming waste stream at a constant flow rate. Various manufacturers use a variety of metering or leveling drum feeders to help prevent surges in the presorting area.

Presorting

The stream of recyclables is fed by the in-feed conveyor to a conveyor line, which delivers materials to a presorting area in the MRF. These areas are designed to further remove contaminants such as stacks of paper, bulky recyclables, and items that could damage downstream equipment. Manual sorters staff work stations in this area alongside the horizontal conveyor. Materials selected are dropped through chutes into roll-off containers or storage bunkers placed directly below the sorting station.

Disk or Star Screens

Early MRF designers incorporated different sorting lines for fiber (OCC, ONP) and containers (plastic and cans) due to their shapes, sizes, and overall density. This design also recognized that the recovery of certain items such as steel or aluminum beverage cans was enhanced when they were not buried under large piles of fiber on the conveyor lines. What has emerged over the last few decades has been the development of specialized sorting equipment to separate fiber from containers.

The emergence of single-stream MRFs having incoming comingled recyclables has required the development of disk or star screens (Figure 4.15), which consist of a series of rotating axles with a number of discus spaced along the axle. The disks (round, oval, or star-shaped) are arranged into rows and desks to form a moving bed. Nearly all of the newly constructed single-stream MRFs in the United States incorporate this kind of screening technology [6].

Disc or star screens allow large materials to travel across the screen while smaller materials (i.e., containers) fall through it. Typically, most MRFs incorporate a series

Figure 4.15 Star screen, Outagamie County single-stream MRF. (From Phillip Stecker)

of disk screens to separate various grades of paper with a primary disc screen used to separate OCC. Industry experts estimate that this type of technology can remove 80–90% of the OCC from the incoming waste stream. The other screens with smaller disks are used to sort other grades of paper. Some MRFs incorporate a secondary screen, called a "polishing screen" to separate the remaining mixed paper, containers, and other residual materials.

Sorting Lines

Once the fiber and containers have been sorted into essentially two different streams, they move in separate sorting lines where mechanical automation is employed as well as manual sorting (Figure 4.16). Two basic sorting methods are used in all MRFs:

- Positive sorting: Materials are pulled out of the incoming material mix.
- Negative sorting: Foreign materials and impurities are removed, and the targeted material remains on the conveyor.

MRFs that "negatively sort" fiber have experienced quality-related problems when the materials have been shipped to paper mills. That is, allowing it to run off the end of the conveyor belt after other materials should have been extracted. Positive sorting fiber materials on the fiber line almost always require manual picking of OCC, ONP, and oftentimes high-grade paper. These materials are dropped through metal chutes into storage bunkers below for eventual baling.

Commingled containers are most often processed differently using a positive sorting approach, removing recyclables commodities and then leaving residue materials and other contaminants on the sorting conveyor. Plastics are recovered either using manual sorting or the use of disc screens or air classifiers. Further sorting is accomplished using manual sorting or magnets to recover steel cans, manual separation or

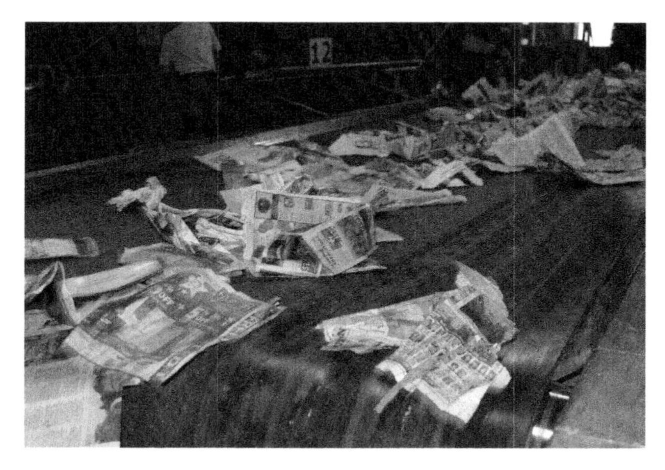

Figure 4.16 Manual sorting line. (From Philip Stecker)

Figure 4.17 Eddy current system.

eddy current systems (Figure 4.17) to recover aluminum cans, glass bottles either by manual sorting or by increasingly common, optical sorting equipment. These materials once recovered are generally dropped through chutes into storage bunkers.

Interim Storage

Most MRF operations employ interim storage bunkers, which are located directly below or near the main sorting line. Typically, these bunkers employ some form of in-floor conveyor to move the stored recovered products from the bunker to baling facilities once sufficient amounts are available for processing (Figure 4.18).

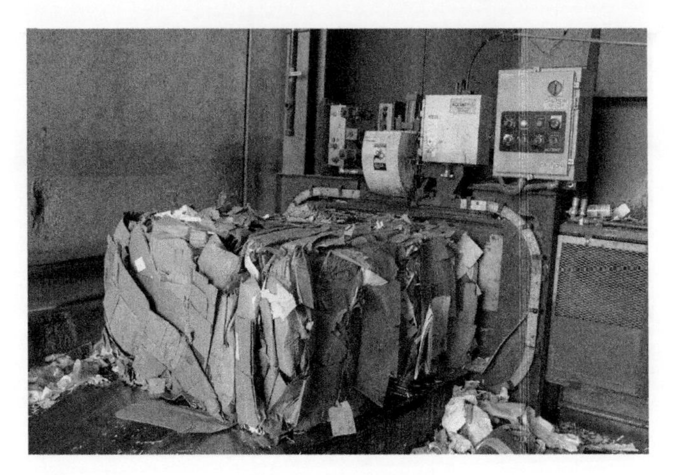

Figure 4.18 Bunker storage.

Consolidation or Densification

The final step in MRF processing is the consolidation or densification of processed recyclable commodities. Typically, most large MRFs in the United States used baling equipment (Figure 4.19) to compress materials into large, dense rectangular cubes that meet market requirements in terms of size, density, and weight. Glass crushers are used in many MRFs to produce a product with constant particle size. Also, many MRFs employ other pieces of equipment such as can flatteners and densifiers for steel and aluminum beverage cans, as well as shredders and granulators for plastic bottles.

Baling equipment is classified as either horizontal or vertical depending on the major direction of the compression ram. Horizontal balers may employ single or two rams, which translates into the number of compressions that the baler employs to produce the bales. A single-ram baler can be adjusted to produce bales of differing size. In comparison, two-ram balers produce a single size of bale but can produce a more compressed bale than a single-ram baler and can be used for a variety of materials.

Specialized Equipment

A variety of specialized equipment (Table 4.6) has been developed for further processing of materials at MRFs. These are briefly discussed in the following paragraphs.

Bag Breakers

Incoming plastic garbage bags pose some challenges for MRF operations. These can be manually slit or can be diverted to automatic bag breakers. Several types of "bag breakers" are available, which typically have specially designed knives installed on the circumference of the drum to open the bags and release the contents. The released materials then are then diverted to the processing line. Other units include heated rods or high-speed rotating teeth.

Figure 4.19 Baling equipment.

Table 4.6 Technologies Used to Separate PET

Technology	Description
Near-infrared	Sensor uses an infrared beam to identify the plastic type by recognizing a light intensity reading unique to each polymer
Laser	Referring to an impurity's spectrum, physical footprint, is able to detect and separate it from the product flow
X-ray	Distinguishes waste based on density,' useful for detecting additives
Color sorting	Separates shades of color seen by the human eye for mixed bottles or flake
Density separation	Flakes sink or float based on relative density to a fluid

Source: From Ref. [7].

Trommels

Trommels are rotating drums used to separate materials by size. These drums have perforations and are set at an angle to enable gravity feed. The rotation of the trommel allows the smaller objects such as pieces of broken glass, small metal particles, and grit to fall through the holes in the perforations. The larger objects work their way down the entire length of the trommel where they exit. MRF designers specify trommels of varying length, which can range from 8 to 80 ft, and varying diameters ranging in size from 2 to 6 ft [6].

Trommels are oftentimes used in mixed waste MRFs to remove fine particles or organics so these materials can be diverted toward a composting operation. Other applications utilize trommels to serve as bag breakers, especially if they are equipped with knives or rods, while other applications employ magnets to help remove metal objects during the tumbling action.

Air Classifiers

Air classifiers are specialized equipment that utilizes blowing air to help separate lighter weight materials such as aluminum cans and plastics from heavier weight materials like glass bottles. This equipment is "chimney like" using a large blower to suck out air at the top of the stack and inducing a high-velocity air stream. The waste stream enters at the middle of the stack, where the lighter weight materials such as office paper and newsprint are extracted by the blower motor. These recyclables are then commonly collected in a cyclone separator that further sorts the materials by size and weight.

A similar type of equipment is the air clarifier, which utilizes a more precise air knife to produce high-velocity airflows working in parallel to help sort the incoming materials. This type of configuration prevents swirling and remixing of the extracted materials enabling further separation of materials such as various grades of paper that differ slightly in terms of mass and density [6].

Eddy Current

An eddy current separator use magnets to remove nonferrous metals, primarily aluminum cans, from the incoming recyclable stream in an MRF. This equipment works by enabling magnetic rotors to spin rapidly to help develop an electric field in the nonferrous metal as it moves across the conveyor line. The nonferrous metal develops an opposite polarity to that developed by the rotor. Consequently, the nonferrous metal is then repelled away from the rotors.

Optical Sorters

Currently, two types of optical sorting technologies are used to sort plastics by resin and color. In the spectroscopy equipment, light waves are emitted whereby each type of plastic on the sorting line reflected back a unique wavelength (Figure 4.20). A sensor installed on the equipment then decides how to classify the plastic into a separate category. In comparison, X-ray technology is used instead to identify the elemental form of the plastic resin. In the color separation technology, a variety of

Figure 4.20 Optical sorters. (From Phillip Stecker)

different camera equipments are used to differentiate slight variations in the color of the plastic to help separate the plastic streams.

One of the most common types of optical sorting technology used for glass separation is light spectro-photometry, which can distinguish between various colors of glass and ceramics. This equipment uses the wavelengths of the different colors to trigger a near-infrared sensor and tell the sensor what color the glass is that is passing by. This in turn helps trigger an air blower that shoots a stream of air at the glass pushing it into the appropriate sorting bin. Manufacturers of these types of equipment claim sorting efficiency of 90–95% [7].

Types of Technologies

A "clean MRF" refers to an MRF that accepts recyclable commingled materials that have already been separated at the source from MSW generated by either residential or commercial sources. There are a variety of clean MRFs, with the most common being single stream where all recyclable material is mixed, or dual stream, where source-separated recyclables are delivered in a mixed container stream. With the advancement of automated single-stream MRFs and the increasing sophistication of new material separation equipment, modern single stream facilities are "state of the art" in terms of use of technology and ability to achieve end product quality that is acceptable to most product buyers. The worldwide market for recycled materials is continuing to evolve, and is expected to remain subject to variability in economic conditions generally, while offering opportunities for refinement and diversification of the materials that are separated. Accordingly, markets will be driven by new technical advances and ability to provide better quality of separation, which in turn will induce equipment suppliers and MRF operators to provide better equipment as prices and demand dictate.

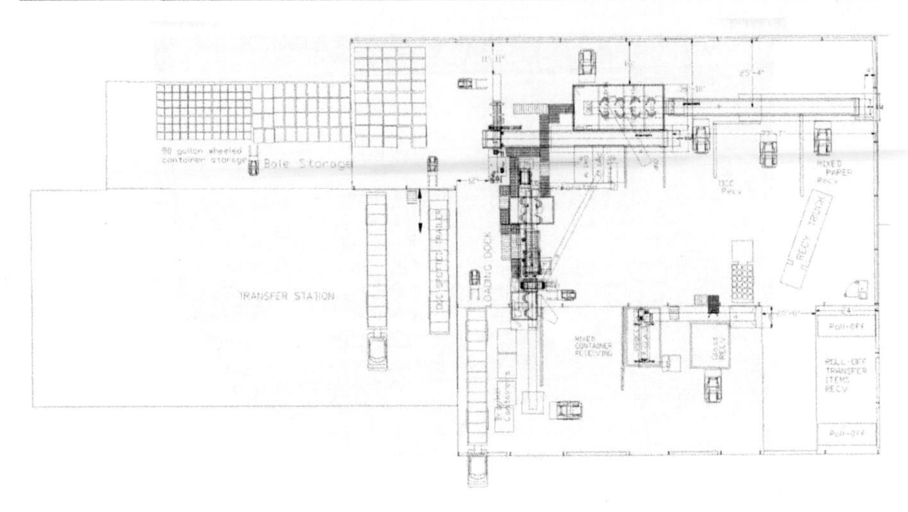

Figure 4.21 Processing schematic of a dual-stream MRF. (From SCS Engineers)

The nation is currently trending toward single-stream recycling collection and processing facilities. Every week news articles can be found reporting a municipality or county making the switch from source separated to single stream. However, it is important to realize the role population plays in the decision to go source-separated as opposed to single-stream: the larger a community, the more they stand to gain from going to single stream. The advantage of single stream is that although source separated has the potential to generate more income per ton due to cleaner material, the sheer volume increase associated with single-stream creates an overall economic advantage for this system. Therefore, the smaller a community, the less they have to gain from volumetric increases, and thus may choose to rely on cleaner, source-separated material to gain an economic edge.

Dual-Stream Processing Facility

In this type of MRF facility, the incoming materials are received by the facility in two streams: fiber (newspaper, corrugated cardboard, mixed paper, magazines, etc.) and commingled containers (plastic, glass, metals, and increasingly aseptic containers). Figure 4.21 shows an illustrative processing schematic of a dual-stream MRF.

Single-Stream Processing Facility

In comparison to the dual-stream MRF, incoming materials for processing are received in a "single-stream." That is, having fiber and commingled containers combined. These types of MRFs include processing equipment that separates these streams of materials into two streams (fiber and containers), which are then further processed in the facility similar to that of the dual-stream MRF. Figure 4.22 shows an illustrative processing schematic of a single-stream MRF.

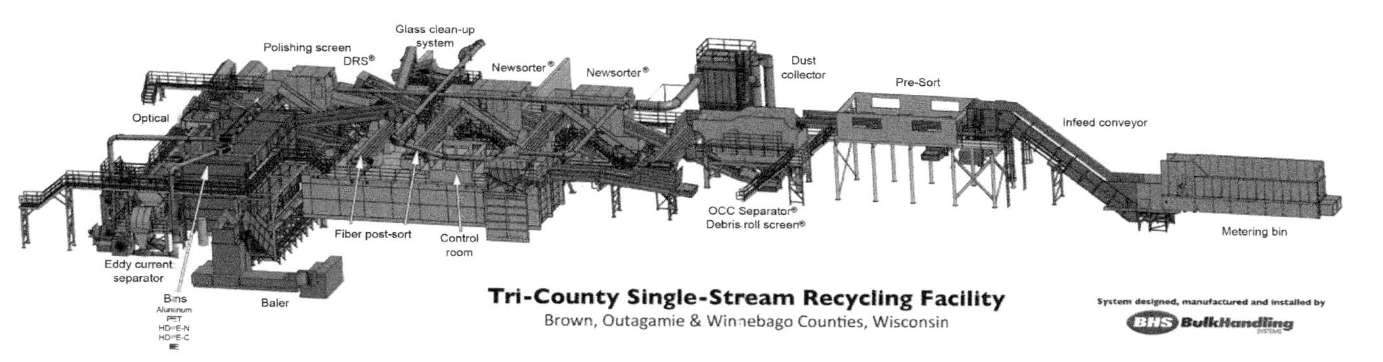

Figure 4.22 Processing schematic of a single-stream MRF, Outagamie, WI. (From Bulk Handling Systems, Inc)

Mixed Waste Processing Facility

Mixed Waste Processing Facilities (also referred to as a "dirty" MRF") receive mixed solid waste (meaning recyclable and nonrecyclable materials, unseparated) which is sorted to separate recyclable material that is then processed (Figure 4.23). Because Mixed Waste Processing Facilities accept one unsorted stream of waste and recyclable materials, they potentially allow for lower collection costs.

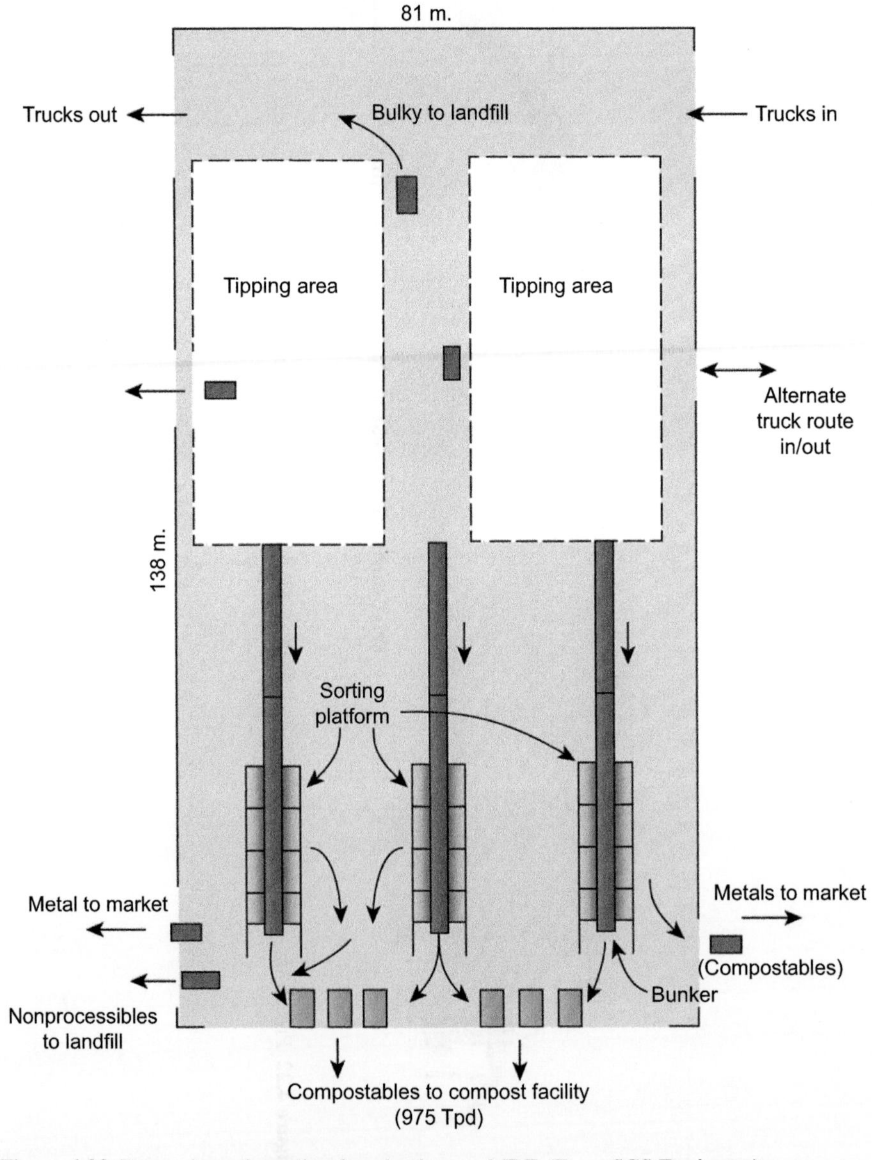

Figure 4.23 Processing schematic of a mixed waste MRF. (From SCS Engineers)

Capital and operating costs are typically higher than a conventional MRF due to the need for more extensive sorting equipment and labor [8]. The potential for contamination is higher; resulting in lower quality recovered materials, as well as lower recovery rates, which can contribute to lower revenue from recyclable material sales. Mixed Waste Processing Facilities are able to achieve recovery rates of 45–70% of the incoming waste as recyclable and compostable materials.

In recent years, communities are re-evaluating the development of a mixed waste MRFs because these facilities can potentially allow processing of waste streams from underserved waste generators such as multifamily and commercial businesses and increasing diversion from the landfill of potentially recyclable materials. Some of these newer mixed waste MRF facilities have been colocated at community transfer stations or landfills to tap into recyclable-rich waste streams. Other communities are evaluating the feasibility of producing high calorific, refuse-derived fuel (RDF) as a means of reducing coal use in certain industries.

The development of mixed waste MRFs is still controversial in the United States. Many still argue that sending a community's entire waste stream without source separation lacks important public responsibility. However, with increasing diversion goals being mandated, many solid waste agencies seem compelled to re-evaluate these mixed waste facilities (Table 4.7).

Table 4.7 Differences Among MRFs in the United States

Criteria	Dual Stream	Single Stream	Mixed Waste
Incoming waste stream	Commingled containers and mixed fibers in separate streams	Commingled containers and mixed fibers in one stream; glass may be separate	Recyclables mixed with nonrecyclables, preferably with organics and wet waste removed
Estimated percentage of MRF systems	52%	33%	<5%
Average residue levels	With glass: 6.7% Without glass: 5.84%	With glass: 11.71% Without glass: 8.1%	Range: 25–75%
Average throughput per processing line	137 Tpd	206 Tpd	400–2400 Tpd
Specialized equipment	Standard MRF equipment	Inclined disk screens to separate fibers from containers; polishing screen	Bag breaker; drum separator, trammel, and/or vibrating screen to separate recyclables from MSW
Final product quality	Typically high with minimal contamination	Increased risk of cross-contamination between containers and fiber	Variable depending on feedstock and processing line
Average facility size (square feet)	10,000–50,000	50,000–150,000	50,000–200,000

Source: From Ref. [6].

Trends in MRF Design

MRF design has evolved over the past several decades due to a number of important drivers, principally changes in governmental policies and expanding recyclables markets. The following major trends in MRF design are outgrowths of these developments. These are briefly discussed in the paragraphs below.

Single Stream

The trend away from dual stream in favor of single-stream recycling is rapidly turning into a major movement across solid waste agencies in the United States. "Single-stream" recycling refers to a collection method in which all unsorted or commingled recyclable materials are collected in one container at the curb and placed in the collection vehicle in a commingled state until processed at a specially designed material recovery facility ("single-stream" material recovery facility). As shown in Figure 4.24, single-stream recycling has experienced a tremendous growth in the United States over the past decade. It is estimated that single-stream MRFs represent more than 40% of all the MRFs operating in the United States as of this date. Nearly half of these units are located in states west of Mississippi due to more stringent recycling goals.

Increased Automation

With the increasing trend away from curbside source separation and the increasing numbers of recycling programs, MRFs have become more highly automated as well

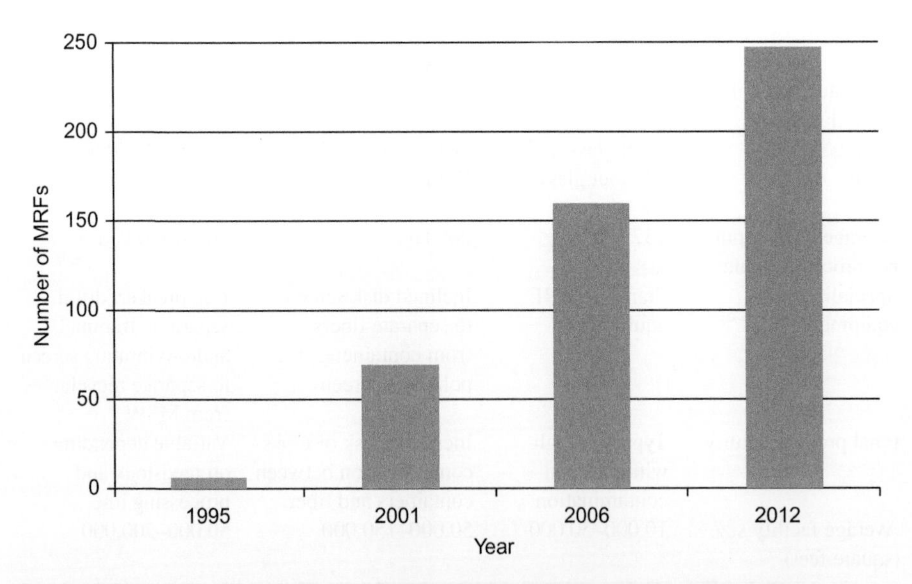

Figure 4.24 Number of single-stream MRFs in the United States. (From Eileen Berenyi)

as increasing in design throughput capacity. Recent surveys of the recycling industry have shown more reported application of optical scanners, drum and eddy current separators, and air classifiers, as well as increasing retrofits of dual-stream systems to handle single-stream recyclables [7]. This application of technology has resulted in a reduction of manual sorting labor on the picking lines, although manual sorters appear necessary in many facilities to ensure quality control over the recovered products.

Larger MRFs

Figure 4.25 shows the trend in facility daily processed per day, which has increased significantly over the past 20 years. These data show that operators are designing MRF facilities with larger throughput capacities in mind to help take advantage of plant economies of scale and pay for the costs of automation. Facilities are becoming more "regional" in nature with recyclables oftentimes being shipped by long-haul transfer trucks or marine barges via "trans-shipment stations" to optimize the cost of long distance transportation. The increasing market area for recyclables is necessary to pay for the increasing capital of these larger and more automated MRFs.

The increasing cost of these plants and regionalization has resulted in a sea change of plant ownership. As of this writing, only 20% of MRFs in the United States are publicly owned and operated. It appears that the difficulty to plan, finance, and operate these increasingly complex plants has shifted ownership and operations of MRFs to private sector companies.

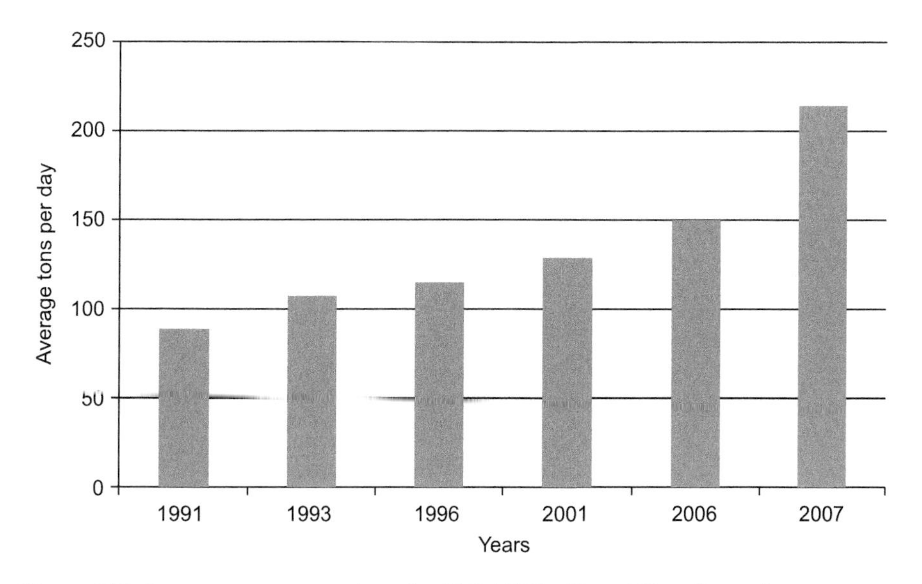

Figure 4.25 Average tons processed per day, U.S. MRFs. (From Eileen Berenyi)

Mixed Waste MRFs

The primary role of this "mixed waste MRF" is to recover materials that would otherwise be landfilled, with the residue from this facility being landfilled. The solid waste industry commonly defines the term "mixed waste MRF" as a facility that accepts loads of mixed waste for the purpose of separating and diverting recyclable materials or organics from the waste stream and transferring the remaining waste for disposal.

Mixed waste MRFs, previously known as "dirty" MRFs, were initially developed in some communities during the early 1990s. Industry experience suggests that most facilities achieved low material recovery rates (5–45%), had high residual rates, and experienced numerous operational and financial challenges [8]. For these many reasons, this method was largely abandoned for construction of new facilities during the 1990s. Today, one authoritative source suggests that these mixed waste MRFs represent less than 5% of the number of operating MRFSs in the United States [6].

In the past 5 years, however, a number of factors have combined to reinvigorate efforts to develop and operate mixed waste MRFs, including high energy costs, aggressive waste diversion goals, favorable commodity values, rising tip fees, and technological advancements in separation equipment. Several new facilities in California (i.e., Athens, Lincoln, and Sunnyvale) have been constructed or retrofitted with new mechanical sorting equipment. In March 2013, the City of Houston was selected as one of five winners of Bloomberg Philanthropies' Mayors Challenge, a competition designed to spur innovation in America's cities. The City won a runner-up prize of $1 million with its proposal "One Bin for All" to allow residents to mix trash, recyclables, and lawn waste in a single bin ready for automatic sorting.

Green Building Design

Increasingly, the newest MRFs are incorporating sustainable or "green building" standards in their design and operation:

- Certified wood from responsibly managed forests.
- Building materials containing recycled content.
- Materials with zero or low amounts of volatile organic compounds.
- Energy efficient components.
- Water conserving fixtures.

Benefits of green design include: federal or state incentives and grant money, lowering operating costs, and improved public and political perception. LEED (Leadership in Energy and Environmental Design) certification, which was designed by the U.S. Green Building Council (USGBC), provides a way to verify sustainability efforts. Platinum is the highest designation given by the USGBC.

The Shoreline Recycling and Transfer Station, which is located in King County, Washington (Seattle), was awarded Platinum LEED Certification in 2008. The facility's design incorporates many "green elements" such as: rooftop rainwater harvesting system, solar electricity panels, skylights, passive ventilation, and recycled building components.

Meeting these types of green building standards provides some added benefits such as reduced operating costs but can add capital costs to these facilities. Some designers have suggested that these features add about 2–8% to achieve LEED certification [6]. Consequently, some MRF owners have applied green building design principals but have not pursued certification.

Construction and Demolition Debris Recycling

Background

Construction and demolition debris (C&DD) is the material generated from the demolition of buildings and other similar civil works infrastructure, the vegetation from land clearing including rocks and soil, and the residual material from the construction of a structure. Thus, specific types of materials could include:

- Metals—reinforcing steel, steel shapes, wire, cans, conduit, etc.
- Concrete, brick, stone, tile, rocks, and soil
- Plastics—bottles, buckets, containers, drums, wrapping, etc.
- Rope
- Paper—cardboard
- Wood and dimensional lumber, pallets
- Ceramics
- Gypsum drywall
- Plaster
- Asphalt
- Rubber
- Foam—insulation, packing, etc.
- Adhesive, glues, paint, etc.
- Vegetation.

C&DD waste presents unique recycling challenges because of its nature. It is inherently a relatively dry waste that when agitated and processed produces a lot of dust. Materials can be combined and intertwined in a way that makes separating them difficult. Most of the material is bulky, heavy, and/or abrasive and results in extensive wear and tear on machinery and can also be a safety hazard to plant workers if proper controls are not observed.

Recycling of parts of demolished buildings has been conducted since the turn of the twentieth century, for example, the reuse of steel beams and columns recycled into automobile bodies and reuse of bricks. However, several factors that emerged in the 1980s have stimulated the growth of C&DD recycling into a large and varied industry sector encompassing many other materials. These factors include:

- Sustained cycles of significant new construction in many major cities.
- The development of relatively inexpensive C&DD landfills to take residual material leftover from the recycling process.
- The emergence of other markets for recycled materials, as a result of the increased use of alternative feedstock for energy production.

- The development and refinement of automated machinery to process C&DD derived from municipal waste streams.
- Development of formal "green" building development initiatives that advocate recycling of municipal waste.

Most C&DD recycling operations in the past relied mostly on manual labor to sort materials. The current generation of these recycling facilities can consistently remove as much as 90%, or more, of the recyclable material from the waste stream, using mostly automated machinery, resulting in a very low production of residual material leftover for landfilling (Figure 4.26). Manual labor is still employed, but for specific tasks where the benefit/cost for replacing a worker with a machine that can be as effective is not economically attractive.

Facility Siting

The location of a C&DD recycling facility must take into consideration the potential for the facility to be a significant additional source of heavy truck traffic, noise, odor, wastewater (from roll-off cleaning), and dust and for these to potentially be a nuisance to the surrounding area if not properly managed. Municipal zoning laws typically restrict these types of facilities to light and heavy industrial districts where these factors typically result in less impact to surrounding properties, however, that does not mean they all become non-issues. Dust control is probably the biggest challenge to C&DD operators and still may be a problem in these industrial districts.

Being a major solid waste management facility, many of the basic principles that apply to properly siting of other major waste management facilities (i.e., transfer stations, landfills, composting) will apply as well. In addition, because the public and

Figure 4.26 Construction demolition debris recycling plant. (From Keith Howard)

other businesses can potentially be adversely impacted by the new facility, they can be expected to be actively involved, and potentially influential in the facility siting process. The siting of a facility should be a carefully planned process that in addition to assessing all of the "technical" and environmental factors, likely will require engaging the public and other businesses early in the process, to alleviate concerns, garner their support, so that the project can move forward to fruition.

C&DD Waste Processing/Recycling Operations

Most of the automated machinery used in current C&DD recycling plants has been derived from the basic materials mining industry and modified for the commercial market. In some cases, new equipment has been designed specifically for municipal C&DD materials. The planning and design of a recycling plant is based on many factors that include:

- Volume of C&DD material projected to be managed over the next 5–10 years.
- The reuse markets that exist or can be developed for recycled materials and the revenue potential.
- Location, size, and availability of property relative to the market being served.
- Development and operating budget (this includes the use of manual labor and automated equipment).

The primary goal in a C&DD recycling facility is to be highly efficient, meaning extract as much of the valuable material as possible while producing the lowest amount of residual material. For a privately owned recycling business this is strictly an economic issue. When the total cost of processing a C&DD waste to remove a specific material exceeds the revenue generated from the sale of that material to the reuse market, then the material is not valuable from a recycling standpoint. Accordingly, the disposal of residual material in the landfill is a cost to the plant operator so reducing residual disposal costs enhances the net revenue from the sale of recyclable materials.

The economic "equation" for a municipally owned/operated facility is a little different, in that this would be a public service and demonstrating a healthy "profit" is not the prime focus. However, the process efficiency goal is still the same whether the facility is private or publicly owned.

What is determined to be a valuable material can change over time as the market, and more specifically the price paid changes for various materials. Thus, the recycling plant must have some flexibility built-in to the separation process and the overall "floor plan layout" in order to be able to process a variety of valuable material and be responsive to changing markets

Processing Schemes

Design of a processing system for C&DD waste follows an approach that, in general, encompasses the following basic principles listed below, starting at the waste tipping area, progressing through mechanized and manual separation then to the temporary storage areas for each targeted material. Some or all of these may be incorporated depending on the factors previously mentioned. Keep in mind there are many different technologies and equipment that can accomplish your goals and we could not cover them all. We have only described herein a few proven techniques.

A schematic illustrating the basic material separation and flow processes is provided in Figure 4.27. Following along with that exhibit, the descriptions below correspond to the letters designating certain features on the exhibit;

- Initial agitation to start to separate the mix of material to improve productivity of downstream processing operations (B).
- Separation of smaller material from larger material. This split usually is around an approximate diameter of 8–12 in. (20–30 cm) (B).
- Separation of ferrous metal to prevent damage to downstream equipment (D).
- Separation of small stones, glass pieces, sand, and grit to reduce wear and tear on downstream equipment (E).
- Separation of light, "two-dimensional" materials (i.e., paper, cardboard, and plastic film products) from smaller and/or denser "three-dimensional" materials (i.e., small containers, bottles, other objects, pieces of nonferrous metal, and sticks of wood) (F).
- Separation of wood/lumber, either with optical machinery (J) or water bath.
- Separation of nonferrous metal, other wood, concrete pieces, asphalt, drywall, etc. This is usually done manually on a "picking line" conveyor system (described below) (C, G, and I).

Sorting and Processing Equipment

Bulk Material Recycling

The recycling of large pieces of concrete, steel-reinforced concrete, and asphalt is a process that requires specially designed heavy duty machines. Concrete and asphalt recycling is done either at a permanent site or at a mobile processing system that can travel to the specific site if there is an extensive volume of material and bulky and heavy pieces that would be cost prohibitive to transport to the recycler. These machines can remove and recycle the steel reinforcing bars and crush the remaining concrete and aggregate to specific sizes. A bulk operation may be separate from an operation receiving a mixed waste stream with smaller pieces because of the land area required for the storage piles at the former operations.

A somewhat unique activity that is common in south Florida is the filling of rock-pit lakes with certain C&DD material considered to be "clean debris." Clean debris is clean concrete, brick, soil, and rocks. Rock-pit lakes are very common in the region and are excavated to obtain the limestone which is a popular and inexpensive source for concrete aggregate and for use in road subgrade. Later when the quarry operation moves away, some of the lakes are back filled with clean debris to make buildable ground.

This section will focus more on the mixed waste operations.

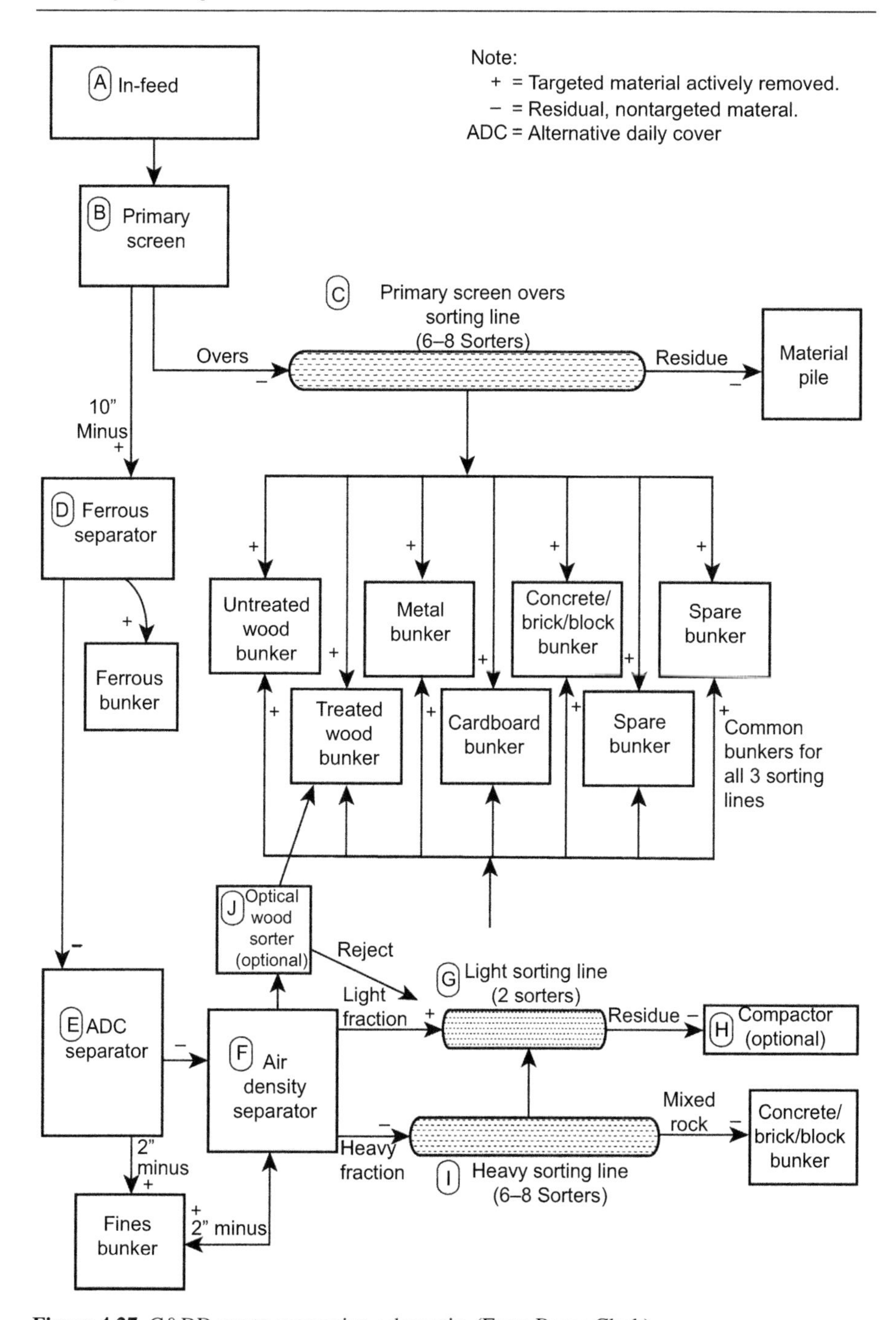

Figure 4.27 C&DD waste processing schematic. (From Bruce Clark)

Mixed Waste Recycling

Screening Operations

Finger Screen C&DD waste is often a clump of different size and types of materials intertwined together. These clumps must be separated at the front end of the process in order to make the separation and removal of the specific material as efficient as possible in downstream operations and to reduce the potential for downstream machinery to become clogged.

A popular and an effective device known as a *finger screen* is often used for this purpose. A finger screen is a heavily built mechanized conveyor that utilizes an articulated, oscillating floor to break up the waste clumps. The waste material is fed into one end and the combination of oscillating motion and the downward slope of the conveyor induces the waste to move forward and at the same time start to separate (Figure 4.28).

At the downstream end of the screen another feature can provide a rough separation of waste by size and density. If this division is desired, the conveyor floor has an opening in which smaller and denser materials (i.e., stones, wood and metal pieces) fall through the opening and are dropped onto another conveyor that takes them to another sorting process. Lighter, flatter, and/or bulkier materials (i.e., cardboard, paper, plywood, plastic film) are carried off the end of the screen conveyor and into the next piece of sorting equipment.

Trommel Screen A *trommel screen* can be used to accept the material from a finger screen. If no size and density separation occurred on the finger screen, then the trommel can be used for this purpose. A trommel is a device that contains a hollow perforated tube, through which the waste moves. The trommel tube is perforated along its length and its circumference with holes or perforations of a specific diameter.

Figure 4.28 Vibrating finger screen. (From Keith Howard)

The tube is angled similarly to the finger screen and is slowly turned on rollers by a motor. As the waste moves through the tube, the rotating action and downward slope move the materials forward where they encounter the perforations. Pieces of waste smaller than the perforations (known as "unders") will fall through the holes and onto a conveyor that will take that material to another sorting process. The larger pieces of material (known as the "overs") will move through the trommel and onto the next sorting process.

Star Gear Screen Another mechanized device used to separate waste into two size fractions is a *star gear screen.* A star gear is a conveyor that contains multiple rows of gears affixed to rotating axles. Each axle contains multiple gears which are spaced at precise intervals along the axle. The spacing of the gears is designed to affect the separation of material. As waste material moves into the upstream end of the screen, it is propelled forward by the rotating gear assemblies. As the material moves over the gears, pieces that are dense and smaller than the opening between the gears fall though opening. This may include stones, pieces of wood, metal, plastic, brick, and similar material. Those pieces (the "unders") are collected by another conveyor and taken for further sorting or disposal. Pieces that are lighter and larger than the opening (the "overs") continue to move along the conveyor, essentially floating over the gears to the downstream end where they flow into the next process. Overs may include cardboard, other papers, pieces of lumber, drywall, plastic film, and similar materials.

Ferrous Metal Separation and Removal

If large amounts of heavy ferrous metal objects are anticipated in the waste stream, then a device known as an *overhead magnet* is used to remove them from the other wastes. The device is essentially a large industrial magnet surrounded by a continuous looped conveyor belt. The entire device is suspended on a structural frame positioned perpendicular (at a 90° angle) above a conventional conveyor belt. Ferrous objects on the conventional conveyor passing under the energized magnet are pulled from the belt onto the magnet's moving belt and are swept away from the lower conveyor. As the looped belt passes out of the magnetized zone, the object is released and falls into a container for removal.

Other configurations of magnets are available for removal of smaller and lighter types of ferrous waste such as metal cans. These magnets, known as *pulley head magnets*, can be located within the end rotating pulley system of a conventional conveyor belt system. These magnets operate on a similar principle to the overhead magnet. Cans on the conveyor belt passing over the magnet are retained on the belt as it travels around the pulley to a point almost directly under the belt where the magnetic field ends. At that point the can falls off the belt and into a container. Nonferrous material will flow off the end of the belt and with the help of the belts momentum, will take a trajectory away from the conveyor into a separate container.

Heavy and Light Material Separation

De-Stoner/Air Knife Separation of light, "two-dimensional" materials such as paper, textile plastic film, aluminum, and pieces of cardboard, from heavier, bulkier

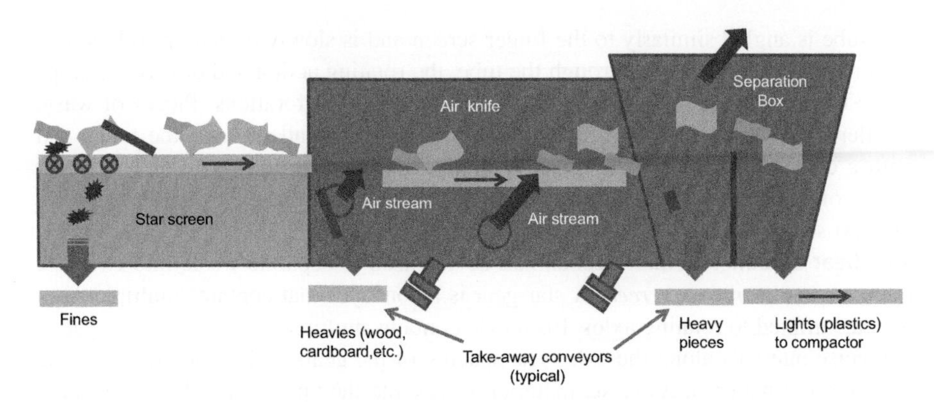

Figure 4.29 Cross section of de-stoner/air knife. (From Bruce Clark)

materials can be accomplished in an automated device called a *de-stoner/air knife*. Figure 4.29 shows a schematic cross section through a unit with two air knifes. This unit uses directional air currents and a vibratory motion to stratify and separate lighter material from heavier material. The unit is mounted on heavy coil springs to reduce the transmission of vibration to the ground.

Following along on this exhibit the waste is fed in from a star gear screen and immediately encounters a gap in the air knife through which high velocity but low-pressure air is flowing. The airflow is provided by a standard centrifugal blower. This flow of air blows the lighter and two-dimensional materials up and toward the downstream conveyor. The heavier and bulkier materials, including small stones and pieces of glass and metal, are unaffected by the air current and fall through the gap onto a take-away conveyor. Thus, the designation as a "de-stoner."

As the lighter materials are conveyed to the end of the air knife, there is another high-velocity, low-pressure air stream directed through the conveyor. This final current of air separates the very light material (mostly plastics and light paper) from other denser bulkier material. The lighter material with a relatively large cross-sectional area is carried to the far end of the collection bin, while the denser, more compact material does not travel as far and drops into the bin directly at the end of the conveyor. All air is exhausted out the top of the air box. If a hood is not used over the final air discharge to capture the light product, then a grate or similar screen is used to deflect light material into the air box end bin.

Exhaust air can be captured and rerouted back to the blowers to increase efficiency and reduce discharge of dust to the near environment. This feature is recommended when the unit is used inside a building and outside where migrating dust could pose a nuisance to other operations or adjacent businesses.

Optical Sorter The use of automated machinery employing "electronic eyes" to assist in separating materials has become increasingly popular (Figure 4.30). In many cases, the increased efficiency and higher purity targeted material provides a

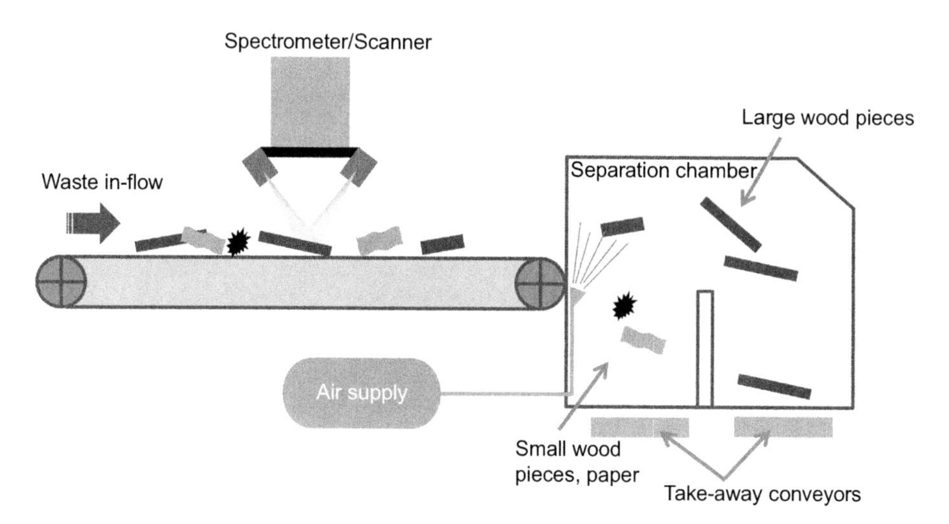

Figure 4.30 Cross section of optical sorter. (From Bruce Clark)

positive benefit/cost ratio as compared to traditional manual methods. The function of the equipment is based on the principle that all solid materials have a unique surface "signature," that reflects and absorbs light rays in varying amounts.

Figure 4.30 shows a general cross section of an *optical sorter* than can be used to separate out large wood pieces.

Following along on this exhibit, as the waste material enters from the conveyor and passes under the control unit, a bright light illuminates the materials. A sophisticated instrument called a spectrometer imbedded in these machines "reads" the reflected light from the materials and through a computerized interface tuned to see wood product, actuates a compressed air device which sends a blast of air that is channeled by the computer program to specific multiple ports positioned across the end of the conveyor belt. As the materials pass over the ports, the ports that have been activated will discharge a blast of air under the material. The air ejects larger wood pieces to a receiving hopper on the far end of the machine, while nonwood material simply rolls off the end of the conveyor belt into a separate hopper.

A vibrating pan feed conveyor is often used to feed the optical sorter. This type of conveyor will increase the effectiveness of the optical sorter by flattening out and separating the materials before they enter the electronic eye detection zone of the sorter.

Picking (Sorting) Line

Some separation of materials is still best achieved with manual labor. The "*picking line*" or *sorting line* is a mostly manual work station that is very common in C&DD recycling facilities. Picking lines are used for removal of many potentially valuable

Figure 4.31 Semiautomated picking-sorting line. (From Keith Howard)

recyclable materials including but not limited to: flat pieces of wood, nonferrous metals, asphalt and concrete pieces, and cardboard. A picking line is typically a long and narrow steel work platform elevated at least 10 ft or more above the normal working floor (Figure 4.31). Under the platform, steel or concrete walls form side-by-side bunkers and reused to provide additional platform support. Each bunker is dedicated to temporary storage of a specific material.

One or more conveyors are mounted on the platform, parallel to its length, and workers stand along the length of the conveyor, usually on both sides. The conveyor is set at a fixed height so that workers have a comfortable arm reach to remove material from the moving belt. Conveyor belt widths can vary from 30 in. to approximately 72 in. in width, the maximum practical width that allows a comfortable reach from either side. Workers are typically spaced about 6–8 ft apart. Located between each worker is a steel chute whose bottom is open through the platform floor.

Each worker is assigned to remove (i.e., pick) a specific material. As the conveyor moves material by each worker they pick their targeted material off the belt and drop the material through the chute where the material falls into a storage bunker. This is termed a "positive sort." Any materials that are not picked from the conveyor are assumed to be of no recycling value or not practical to segregate and are allowed to flow off the end of the belt into an end bunker. This is termed the "negative sort" material, or alternately where the contaminant material is selected to be removed from the targeted material.

Periodically the bunkers will be unloaded and the materials taken away for final recycling/reuse or disposal. Bunkers can have mechanized unloading systems or be unloaded with a front-end loader.

Dust Control

Introduction

The processing of C&DD waste typically generates a significant amount of dust. Most of the materials in C&DD are inherently dry and when they are agitated in the various processing operations, dust, actually the minute pieces of some of the materials, are liberated into the air. Dust particles from a C&DD operation can range in size from around 100 (concrete dust) up to 1000 μm (textile dust).

C&DD processing operations outdoors and inside a building can create a nuisance to neighbors with fugitive dust generation if it is not controlled properly. Constant uncontrolled dust can also be a health hazard to workers.

The major sources of dust can include the following areas and pieces of equipment:

- Tipping floor
- Finger screen
- Air knife
- Storage piles (especially for residual "fine" materials)
- Recycled material load-out.

Dust control is mostly science and but also part art, and too extensive a topic to cover completely here. However, fugitive dust is probably the most prevalent problem with C&DD processing. Dust controls can take many forms, to basic static screening material, to misting devices, and/or to mechanized filtering systems. The most common system for waste operations, the misting system is discussed herein.

Wet Systems

Dust is commonly controlled using wet spray systems where enclosing the material area is impractical. Wet systems can be applied for *prevention* and *suppression* of dust. Prevention is applying a wet mist directly to the material to reduce liberation of dust. Suppression is applying a mist to the air around the material once it is agitated and dust is released. The type of misting system depends on the situation. Many C&DD recycling operations may require both. Recycling facilities are particularly sensitive to the correct dust control. If too much water is applied the materials may stick together and significantly reduce the effectiveness of downstream separation activities.

Effective dust control depends on many factors including; dust particle size, wind affects, freezing temperatures nozzle type, nozzle spray pattern, spray locations, available water pressure, use of compressed air, and use of surface wetting agents (surfactants) to name a few. The droplet size produced from the system must be compatible with the dust particle size or removal is reduced. Dust control is best left to companies that are expert in design and operation of these systems in waste handling operations.

Vibration

A brief discussion on vibration is in order. The type of waste processing system described herein contain several pieces of heavy oscillating equipment, as well as a significant amount of rotating machinery, in general. These systems produce steady-state (continuous) vibration and impact (isolated) vibration. Steady-state vibration occurs during normal running operation. Impact vibration can occur when the equipment is started and stopped or an unusually heavy material is dropped into the machinery. The location of such a system should consider the potential for this equipment to transmit these vibrations to the ground, despite the vibration dampening systems designed into most of the equipment. Vibration also results from waste tipping operations and load-out of recyclable materials. A significant buffer zone is generally needed between the equipment and the property line to allow the ground vibration to dissipate to a background level.

Many municipalities have a standard for industrial zones that sets the maximum limit on steady-state and impact vibrations. Without an adequate buffer width, vibration can be transmitted to adjacent properties where it can become a significant annoyance. The distance and strength over which vibration can be transmitted is contingent on many factors including soil type, moisture content, frequency of vibration (i.e., number of oscillations per unit time), and others. It is recommended you consult the equipment manufacturer for guidance on a buffer distance. A buffer distance of 200 ft would be considered a good starting minimum with competent soil, a larger distance is preferred if available.

Recycling By-Products

The processing of C&DD waste with a certain combination of equipment can result in a residual by-product known as recovered screen material (RSM). RSM looks similar to soil and is mostly the combined residual of some actual soil and minute pieces of friable waste that may include: drywall, paint, plaster, asphalt shingle grit, pieces of grout, cement and brick, glass, and plastic.

Some recycling companies have experimented with developing this product for use as a substitute where nonheavy load-bearing natural soil fill is acceptable. For example, in shaping contours on a golf course, filling residential lots up to flood criteria, and similar uses where natural soil is typically used. Although some regulatory agencies have approved its use in certain conditions after an extensive chemical testing program, and it has been marketed with some limited success, extra caution is warranted. Some RSM has found to be contaminated with chemicals that, although they are native to the virgin materials, are of environmental concern because of their potential to leach out of the RSM. These chemicals include polynuclear aromatic hydrocarbons (PAHs), lead, and arsenic.

RSM is ground-up residual from many materials. As a result, the surface area of the individual particles is increased which can increase their solubility when wetted. The increased solubility may result in leaching of the chemical which can result in contamination of the natural soil and groundwater. Common materials containing these contaminants include asphalt roofing shingles (PAHs), paint (lead), and treated wood (arsenic). Removal of the offending material from the waste stream before they go through the separation process so they do not end up in the RSM is a time-consuming and expensive process that requires constant vigil by the operator. It is recommended to steer clear of RSM.

Organics Processing Technologies

As noted in Chapter 3, organics represent roughly 40–60% of the municipal waste stream by weight in the United States. Many of these materials have, up until recently, been disposed of in landfills and WTE facilities. With increasing waste diversion goals or targets in many areas, organics have been viewed as an avenue to help achieve these targets. Further, many provincial governments, states, and local agencies have enacted legislation and regulations to promote organics diversion from solid waste facilities because organics are a major contributor of landfill methane emissions. Many states in the United States have such regulations, which ban or restrict the disposal of yard wastes, a major component of the organics portion of the MSW stream, from landfills [9].

Recent surveys undertaken by the U.S. EPA and *BioCycle* magazine indicate that there are about 3000 yard waste composting programs in the United States [10]. This roughly equates an average of 78,500 Tpd of yard wastes. There is an increasing trend to move beyond yard waste and divert other materials such as food waste, soiled papers, disposable diapers, and pet wastes into composting waste stream. There are currently 150 food waste programs in 16 states [10].

In Europe, organics recovery and processing have been widely practiced for decades. There are over 160 anaerobic digester (AD) facilities in operation, which process more than 11,000 Tpd of organics [11]. There are various reasons why these numbers of facilities have been used in construction, among them the following:

- European Union Landfill Directive: The EU Landfill Directive (Council Directive 99/31/EC) requires that the organics portion of the EU solid waste stream that is landfilled be reduced by 65% in 2016. There are provisions that mandate that organics that do get landfills must be treated. These regulations promote source separation and processing of organics.
- Prices for Renewable Energy: Prices for renewable energy are high in Europe, ranging from $0.15 to $0.25/kW h. Further, there is a regulatory requirement to connect these facilities to the national electric grids as well as providing substantial grants, loans, and other subsidies to promote their construction and operation.
- High Landfill Tariffs of tipping Fees: The EU Directive has had the impact of reducing the numbers of operating landfills and overall landfill airspace available. This has resulted in increased landfill tariffs.

Approaches

Briefly, there are three major approaches to process organics from the MSW stream. These are discussed in the paragraphs below and summarized in Table 4.8:

- Composting the Entire MSW Stream: During the 1980s and 1990s, there were many facilities, which were constructed to process the entire MSW stream [11]. Due to poor quality control and contamination of the resulting compost product, almost all of these closed.
- Processing Mixed Waste Organics: This approach essentially involves the use of recovery facilities to recover recyclables from a mixed waste stream, development of a fuel product (RDF), and an organic fraction that can be composed. While this eliminates the need for separate collection of the individual waste streams, the amount of preprocessing of the incoming waste and further processing of the potentially compostable organics make this approach costly and generally result in a poor compost product that has poor marketability.
- Processing Source-Separated Organics: The source-separated approach mandates the use of separate containers for collection at the source to avoid contamination.

Table 4.8 Methods for Processing Organics from MSW Stream

Approach	Technology	System Requirements			Recovered Products		
		Separate Collection	*Mixed Waste MRF*	*In-Vessel Required*	*Energy Recovery*	*Compost*	*Quality*
All MSW	Composting	No	No	Yes	No	Yes	Poor
	Anaerobic digestion/ composting	No	Yes	Yes	Yes	Yes	Poor
Only MSW organics	Composting	No	Yes	Yes	No	Yes	Poor
	Anaerobic digestion/ composting	No	Yes	Yes	Yes	Yes	Poor
Source-separated organics	Composting	Yes	No	No	No	Yes	Good
	Anaerobic digestion/ composting	Yes	No	Yes	Yes	Yes	Unknown
	Bioreactor cells	Yes	No	No	Yes	Yes	Unknown

Source: From Ref. [11].

Composting involves the aerobic biological decomposition of organic materials to produce a stable, humus-like material. Composting happens naturally in the environment when organic material falls to the soil surface. There are many compost technology options for managing most organic materials in the waste stream, each striving to optimize the biological conditions in the mass of material to achieve the most uniform, mature compost in a reasonable amount of time.

The composting process is somewhat forgiving in practice, so it is not always necessary to meet ideal conditions for making good compost; but, the closer the system can get to the ideal, the better and more consistent the product will be. The resultant compost product makes a valuable soil amendment due to its high organic matter content. Because compost contains high levels of organic carbon, which can fuel key ecosystem functions like nutrient cycling, water retention, and erosion control, it can also help rebuild soils.

Backyard Composting

Backyard composting is a viable method of yard and organic waste reduction for those residences and properties that have space available and are inclined to consider this possibility. Many communities in the United States are encouraging "don't bag it" programs to reduce yard waste collection and processing programs.

This form of composting can be as simple as a backyard pile, which takes a relatively long time to turn to compost, to a tumbler (Figure 4.32) where, if turned everyday could make compost rather quickly. The organics utilized for backyard composting include yard waste as well as kitchen vegetable and fruit scraps.

Individual Composting Units

In recent years, manufacturers have developed self-contained composting units that uses tap water, small wood chips, enzymes and natural, nontoxic bacteria in an

Figure 4.32 Backyard composting unit.

Table 4.9 Comparisons Among Food Digestion Equipment

Requirement/Feature	**Manufacturer**	
	Marathon	***Totally Green***
Food digested automatically	Yes	Yes
Unit capacity	1200 lb (0.6 tons)	800 lbs (0.4 tons)
Minimum cycle time (per unit capacity)	8–12 h	8–12 h
Voltage requirement	208–240/3 phase	110/single phase
Annual power use (based on 8 h/day or 7 days/week)	4350 kW h	3500 kW h
Hot–cold water supplies required	Yes	Yes

Source: From SCS Engineers.

agitated compartment to rapidly decompose food waste into water [12]. The agitator is a paddle that is slowly turned by an electric motor. The wood chips are typically only replenished about every 6 months, while the enzyme/bacteria mix is typically replenished about every 2 months.

Prior to placing food in the digester, a worker charges the digester with the prescribed amount of wood chip and enzyme formula. Then, the food wastes are loaded into the digester. Typically, the digester has a built-in automatic scale and warning light system, that automatically weights the incoming food waste and by illuminating yellow and then a red light tells the operator when filling is almost complete and when the unit is full. The digester automatically adds the prescribed amount of potable water and then starts itself. The food wastes are decomposed by the enzymes and turned into a gray liquid. The process usually takes about 8 h. The gray liquid is automatically flushed from the digester into the plumbing line. The plumbing line shall be connected to drain to a sewer manhole or directly to a sewage lift station. At the end of the day, the sorting pad is rinsed clean with a hose and any remaining food residue flushed down the drain connection. Table 4.9 summarizes the requirements and main features of two digester units on the market in the United States.

Larger-Scale Compost Facilities

Open-Air Windrow Figure 4.33 illustrates a conceptual site plan of an open-air windrow facility. In windrow composting systems, the feedstock is formed into long, narrow piles in which composting takes place. These piles are usually 6–20 ft wide, as high as 12 ft, and generally trapezoidal in cross section. The length can vary, but 300 ft is typical.

There are two different types of designs for these windrow facilities. In the "turned windrow" approach, aeration is supplied by mixing or turning the windrow. In the "aerated static pile" approach, the piles are not moved. Plastic pipes are installed at the bottom of the piles, which are then attached to blowers that introduce air into the piles to improve decomposition.

In small systems, windrow turning is frequently accomplished by using a front-end loader. Specialized windrow turners are used in larger sized facilities to help

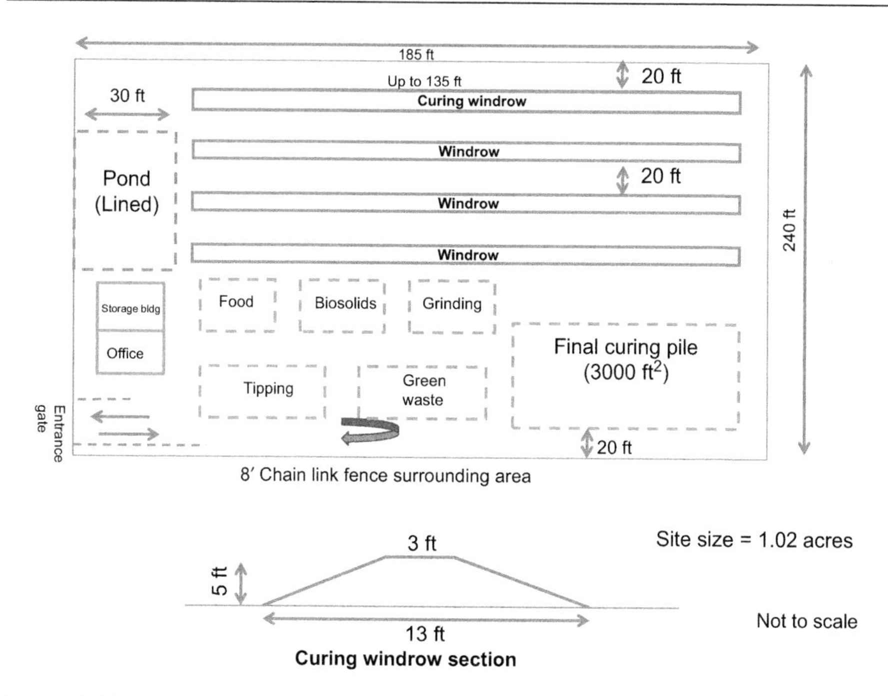

Figure 4.33 Illustrative schematic of a windrow composting facility. (From SCS Engineers)

control the process. Because of the simplicity and low cost, windrow systems have been widely applied for green wastes composting.

To develop the specific area requirements for a windrow composting facility, the volume of yard waste materials and potentially compostable materials in the community's waste stream are calculated. These volumes are then converted to tons delivered to the facility using an appropriate waste density. This calculation determined how many windrows would be needed to accommodate the amount of material received. An appropriate engineering factor is then used by the designer to determine the number of aisles between pairs of windrows, equipment movement space, and ultimately the width and length of the processing area needed. Added to this total is a calculation for the storage and buffer area needed to arrive at the total site area.

Under both approaches, the organic wastes are kept in windrows in the active composting stage for a usual period lasting from 4 to 12 weeks. After this step, the composted materials are allowed to sit in windrows without aeration. This is commonly called the "curing phase," which can range from another 4 to 16 weeks. Composition of the incoming composting waste stream impacts these ranges.

A summary of the pros and cons of the turned windrow and forced aeration approaches is as follows:

1. Windrows are low/no-tech, while forced aeration requires a blower system and personnel to maintain and repair its increasing costs.

2. In turned windrows, the recipe and pile structure can be adjusted after piled, while forced aeration requires proper mixing before placing piles.
3. Windrows can be turned and moved at will, while a forced aeration system must be disassembled before moving materials.
4. Negative pressure-forced aeration can help control odors by collecting air into the suction pipe, enabling filtration before discharging. Windrows require turning to aerate and can release odors as the pile is opened.
5. Positive pressure-forced aeration can eliminate excess moisture and excessively high temperatures by pumping higher volumes of air into the pile than the negative pressure system can pull in. Turned windrows must be turned repeatedly, or mixed with drier materials, to reduce moisture and temperature.
6. During dry weather periods, windrows will hold moisture better than piles processed with forced aeration.
7. Forced air piles can be built as an extended pile, reducing the size of the "footprint" needed to process a given amount of material.
8. Forced air systems need an engineer to design the system to assure the airflow will be sufficient for the amount to be composted.

In the 1990s, stringent odor regulations in Europe have resulted in a trend toward enclosed in-vessel composting systems. For example, the TA Luft (Technical Instructions on Air Quality Control) enacted in Germany essentially demand enclosed composting structures and mandate odor limits of 500 odor unit per cubic meters [11].

In-Vessel Composting Systems In-vessel composting system is a high technology approach consisting of different proprietary systems that usually involve mechanical agitation and forced aeration and may be enclosed in a building. These are the most capital intensive and result in the greatest level of process and odor control, as well as the shortest composting time required. In-vessel systems have been designed with rigid containers consisting of concrete or steel. In recent years, a variety of flexible containers, consisting of flexible membranes, have been invented and applied in many locations.

Flexible In flexible, in-vessel systems, the incoming compostable materials are ground up and then placed in semi-impermeable bags or cover systems. For example, in the GORE Cover System, which is manufactured by W.L. Gore & Associates GmbH, the polytetrafluoroethylene membrane cover is used in combination with a flexible aeration system controlling oxygen and temperatures in the cover. The semi-permeability of the cover membrane to water vapor and air influences the extraction of moisture during composting. This allows CO_2 produced during the composting process to escape. At the same time, the cover works as a physical barrier against odors and other gaseous substances escaping from the composting material. Odors and other gaseous substances dissolve in this film of water and drop back into the composting material where they continue to be further broken down by bacteria. The cover system can achieve a reported reduction of up to 97% in odor concentrations compared with composting in open windrows without aeration control, and this without the need for additional filtering [11].

The integrated aeration system of this technology shortens the composting process. Special probes used help measure the oxygen supply and temperature within the composting material; the measurements are used to regulate both parameters via aerators.

Rigid A rigid, in-vessel composting facility utilizes a mechanized initial decomposition stage to significantly shorten the overall period of decomposition and ultimately the total time required to produce the final product, as compared to windrow composting only. Also, because volume reduction is achieved in this process, a smaller land area for final composting is needed. The final composting period takes less time as well, generally from 3 to 6 weeks.

The mechanization is the use of one or more long steel tubes (in some cases up to 80 m long) that are slowly rotated on fixed rollers with motors. These look very similar to kilns in use at a cement plant. Figure 4.34 is an illustrative concept layout for the in-vessel plant, which was planned for an application in Colombia.

The raw waste stream is initially sent to a prescreening facility module to remove nondecomposable materials. The activities at the preprocessing facility will not be repeated herein. Some in-vessel facilities screen the waste again, at the composting facility, with a trommel to remove some smaller items that may have been missed at the prescreen.

The screened waste is then introduced at one end of the tube which is elevated in relation to the other end and slowly moves by the combination of gravity and paddles inside the tube to the other end of the tube. The inside of the tube is kept aerobic by blowing air continuously through the tube with a mechanical blower. Moisture is added as needed. This operation takes about 3 days and after that time the waste is actively decomposing and breaking down in size.

Material discharging from the tube is sent through two trommel screens in series to remove oversize objects and as much nondecomposable material that is left. The second screening may or may not be necessary but is included in this design. The screens are designed to progressively remove the finer material which is presumed to be the organic materials from the oversize material, which is presumed to be the nondecomposables. The last screen opening would size the material to anything less than approximately 1.2 cm (1/2 in.). The organic screened material will be stored inside a bunker until it is sent to final composting.

Environmental and Public Health Impacts

The proximity of composting facilities to residential, commercial, and institutional facilities has prompted concerns about public health due to the types of airborne particles and odors emitted by these facilities in normal operation. There have been countless studies undertaken, both in the United States and Europe for the EU. The following paragraphs briefly review some of these findings and recommendations.

Bioaerosols

Bioaerosols are defined as particles of microbial, plant, or animal origin and oftentimes are called "organic dust." They can include live or dead bacteria, fungi, viruses, allergens, bacterial endotoxins (components of cell membranes of Gram-negative bacteria), antigens (molecules that can induce an immune response), toxins

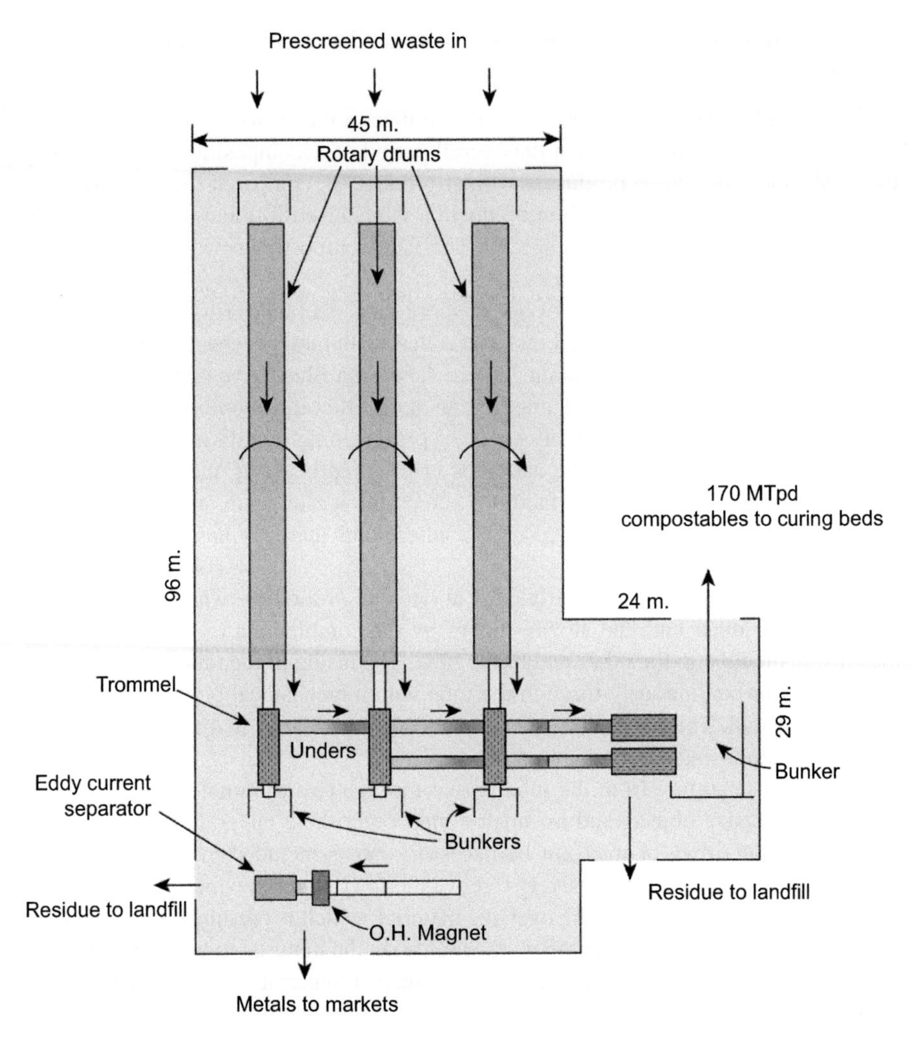

Figure 4.34 In-vessel composting facility. (From Ref. [12])

(toxins produced by microorganisms), mycotoxins (toxins produced by fungi), glucans (components of cell walls of many molds), pollen, and plant fibers [8]. Many of these kinds of bioaerosols are known to cause a variety of human impacts such as infection and sensitivity over even short periods of time. Currently, however, there is no ambient or occupational exposure limits for bioaerosols in the United States. Further, there are no validated standard methods for measuring these bioaerosols.

Odors

As noted above, composting is a biological activity which releases bioaerosols. Odors are caused by these chemical compounds and are generally a nuisance and do not present significant health risk to the community. In general, an "odor" is a sensation our brains generate in response to chemicals in the air which are breathed in through the nose. Humans can perceive most odors even at relatively low concentrations of the chemicals, in parts per billion in many cases.

Briefly, odors are subjective and humans differ in their sensitivity to odorous compounds in the air. Where some people detect an odor to be strong, others sense it as weak or even nonexistent. Further, people differ in their reactions, preferences, and aversions to odors. An objectionable odor to one person can be tolerable to another. The perception of odor depends on both the characteristics and concentration of the odorant and the person smelling it.

Because odors are perceptions of numerous chemical compounds, detecting, measuring, monitoring, and even describing odors are not straightforward processes [13]. The human element cannot be completely removed from the process because odors are perceived subjectively at very low odorant concentrations. Odor science has developed special terms and techniques to describe and measure the nature and strength of specific odors. Both qualitative and quantitative descriptors are necessary—an odor may be measured according to its threshold concentrations, pervasiveness, descriptive quality, degree of pleasant or unpleasantness, and the concentration of the odor-causing chemicals present.

The most common measure of odors is the odor threshold value. Odor strength is quantified by determining the amount of dilution needed to bring the odorous air sample to its threshold of perception. The higher the threshold value, the more dilution necessary to bring the odor to threshold, and thus, the stronger the odor.

The odor threshold is determined by trained human assessors observing presentations of the odorous air sample dynamically diluted with an olfactometer. Because individuals perceive odors differently, odor panels contain several members, preferably 5–10. Their collective response is expressed statistically. For example, odor thresholds are usually defined by the point at which 50% of the panel no longer detects the odor (e.g., D/T50). Results are computed for each assessor based on the dilution levels where correct "detection" or "recognition" responses are recorded. The responses of all assessors are averaged to determine the sample's detection and recognition thresholds.

There are four potential odor sources at an active composting facility:

1. Odors may be present in the material received for processing.
2. Odors may develop from the compostable materials while they are onsite.
3. Compostable materials may be deposited on the ground around the site and develop odors when it become wet and decay.
4. Leachate and storm water on the site may accumulate in puddles, ponds, or tanks and develop odors as the nutrients in the liquid decay.

Compost Management

Given the potential for serious public health impacts from composting facilities, a number of industry experts have suggested the following best practices [8]:

- Siting: Consider how meteorological and topographic features affect airflow; establish buffers between receptors and the composting facility.

- Design: Site material handling processes downwind or maximum distance from receptors; enclose facilities to reduce offsite exposures if needed; install pad type suited to the operation; minimize ponding through site grading and design.
- Operation/Management: Maintain good airflow through the compost; minimize handling; turn compost based on temperatures, not a schedule; restrict material movement to times when the potential for off-site movement is minimal (low wind) and receptor population is least (time of day, avoid weekends and holidays); minimize disturbance of dusty areas by equipment; minimize dust by adding moisture to materials when moving them and to traffic areas and watering dry materials and dusty areas.

Emerging Waste Conversion Technologies

The alternative waste conversion (WC) technologies are numerous and can be grouped many ways; but for this discussion, technologies have been grouped by three major processes that include:

1. Thermal
2. Biological
3. Biochemical.

Within these groups are many methods and technologies that have been developed to extract different benefits from the processed waste stream including:

- Gases for power production
- Gases for feedstock for vehicular fuels
- Basic chemicals for use as a raw feedstock
- Compost/soil amendments
- Slag for use an alternative building material.

A brief description of the main technologies in each of the three groups is presented below with discussion as to potential relevancy to the community and region, benefits, estimated costs, and potential advantages and disadvantages.

Thermal

The thermal technologies are based on taking the solid waste and processing it under moderate to very high temperatures in a closed reactor vessel, sometimes under pressure and with or without the introduction of air or steam. Depending on the particular process, traditional recyclables may be removed at the front end of the process or during the process stages. The predominant processes are pyrolysis–gasification and autoclaving.

Pyrolysis–Gasification

In a pyrolysis process, air is excluded from the reactor vessel and results in the waste decomposing into certain gases (methane, carbon dioxide, and carbon monoxide), liquids (oils/tar), and solid materials (char) [14]. The proportions are determined by operating temperature, pressure, oxygen content, and other conditions. Because there is little to no air or oxygen, the waste does not combust as it breaks down (there are no flames).

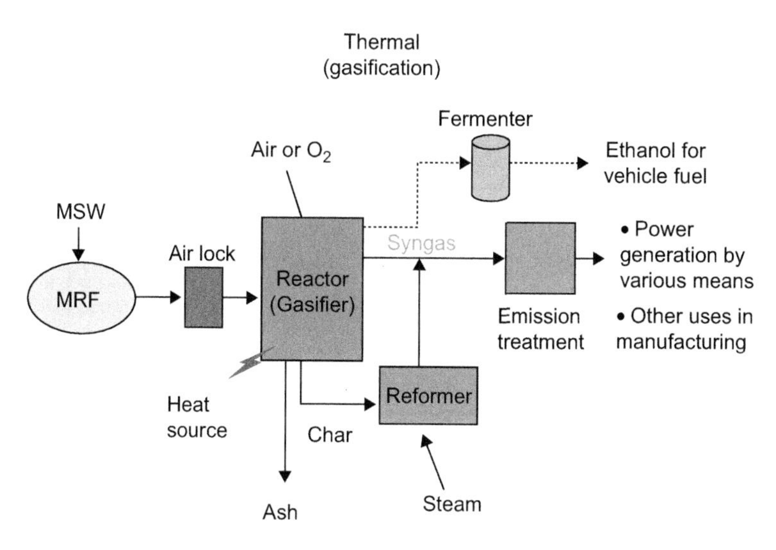

Figure 4.35 Basic gasification process. (From SCS Engineers)

When the amount of air in the process is less than that required to support combustion, but greater than in a pyrolysis process, the process is termed *gasification*. This process is typically used to achieve a different balance of the gaseous by-products, mainly the production of a hydrogen (H)-rich gas with smaller quantities of carbon monoxide (CO), methane (CH_4), and carbon dioxide (CO_2). The refined gas, primarily H and CO, is termed *syngas* and has many direct applications such as powering a turbine to produce electricity and potentially for use as a feedstock to produce alternative vehicular fuel (ethanol) or other chemical compounds. Most of these processes require an external heat source under normal operating conditions. This is usually hot, clean air that captures heat from the downstream gas combustion process.

A basic gasification process is shown in Figure 4.35. Gasification processes have attracted much interest because the process is inherently more efficient than a combustion-based process, the syngas is a relatively clean energy source and the plant may generate less troublesome air emissions overall.

A relatively recent development for solid WC using the gasification process, that employs a unique heating source, is known as a plasma arc converter [15]. Although there are many variations, a typical plasma arc converter uses an array of plasma torches to generate temperatures in the reactor of more than 5000°C. This extremely high temperature, coupled with a gasification environment, has shown potential in small laboratory test units to achieve a very high efficiency in decomposing the organic fraction of the waste to syngas, while generating a slag material from the inert fraction [16]. The slag has potential for use as a substitute ingredient in potentially many building materials, including concrete structural elements (e.g., wall panels and blocks) and asphalt.

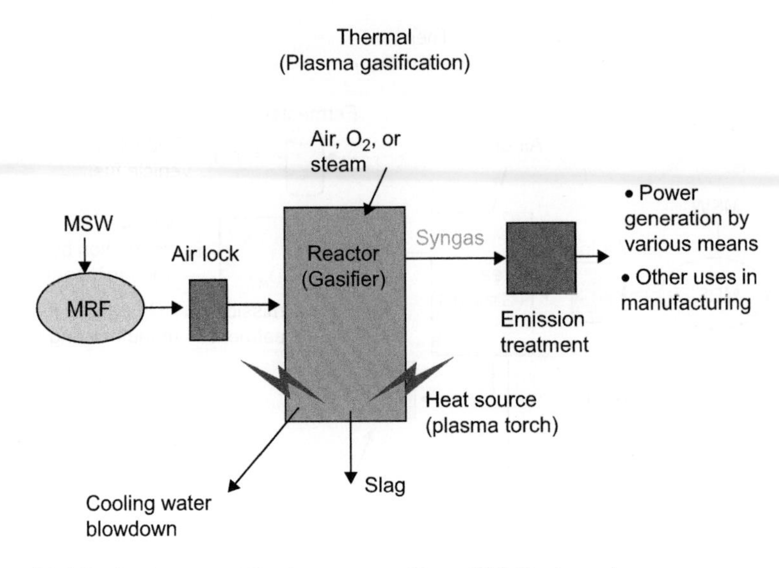

Figure 4.36 Basic plasma gasification process. (From SCS Engineers)

A plasma is an ionized gas that results when a basic gas, such as nitrogen or air is passed through an electrical arc struck between two electrodes. The electrodes are constructed into a torch that directs the plasma arc. The intense heat created by the arc can be used to treat many materials, including MSW. Plasma arcs were commercialized in the metallurgical industry where the high temperatures produced in the reactor vessel (potentially up to 10,000°C) are used to create special alloys [17]. Some of the electric power generated by the plant is siphoned off to power the torches. The basic plasma arc process is shown in Figure 4.36.

Autoclave

The basic autoclave process has been in commercial use for decades, primarily in the medical field for sterilizing instruments, some manufacturing uses, and in the sterilizing of medical wastes [18]. In an autoclave process for solid waste, mixed MSW is fed into a reactor vessel where it is subjected to heat, pressure, and agitation. The reactor conditions cause the organic fraction of the waste (i.e., food scraps, fiber/paper products, and vegetation) to break down into a pulp-like substance that potentially has reuse applications depending on the degree of postprocessing selected.

The pulp has been demonstrated with a few systems to be a useful soil conditioner and also is being tested for use as feedstock for the production of ethanol, an alternative vehicle fuel and in the production of an RDF for combustion in power plants. The process also claims to provide a higher quality recyclable product. Plastic recyclable materials are softened and occupy less volume downstream. Product labels on glass, plastics, and metals are totally removed, and these materials also are cleaned and sterilized. A basic autoclave process is shown in Figure 4.37.

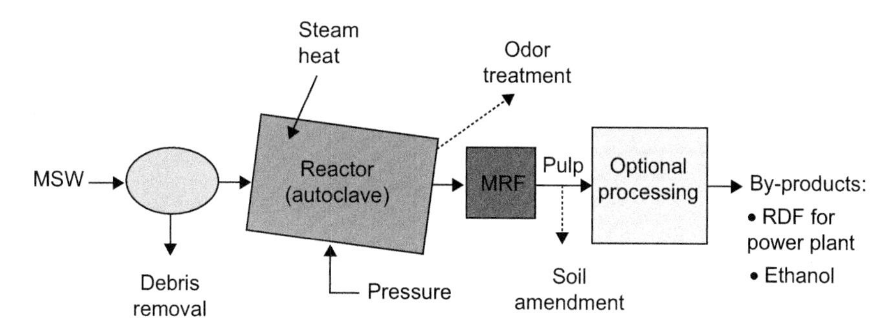

Figure 4.37 Basic autoclave process. (From SCS Engineers)

Biological

There are two types of biological processes being utilized for WC. These include the anaerobic and aerobic process technologies. The following paragraphs briefly describe these technologies.

Anaerobic Process

Anaerobic digestion is the bacterial breakdown of organic materials in the absence of oxygen. This biological process produces a gas, sometimes called biogas, principally composed of methane and carbon dioxide. The anaerobic process is often used to treat organic wastes other than nonsegregated MSW, and that is where it is used the most. This anaerobic process is used to digest sewage sludge (i.e., biosolids—produced from treated sanitary sewage), yard vegetation, agricultural wastes (both animal and plant), and some industrial waste sludge. The number of plants processing these materials is currently in the thousands worldwide.

The anaerobic digestion process occurs in three steps [15]:

1. Decomposition of plant or animal matter by bacteria into molecules such as sugar.
2. Conversion of decomposed matter to organic acids.
3. Organic acid conversion to methane gas.

Depending on the waste feedstock and the system design, biogas is typically 55–75% methane. A basic anaerobic process is shown in Figure 4.38.

Aerobic Process

The aerobic process relies on a continuous supply of air to be mixed in with the waste material. Again, the waste is ground up into pieces. Recyclable materials are removed before this process. In a typical plant, the waste is ground up and formed on an outdoor pad into long piles called windrows. The windrows are agitated a few

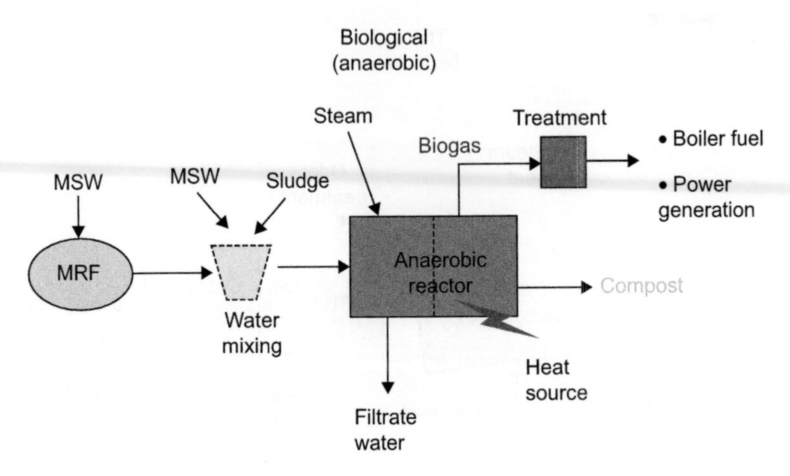

Figure 4.38 Basic anaerobic process. (From SCS Engineers)

times per week to allow all parts of the pile to be exposed to air. The agitation and aerating process can also be conducted in a vessel into which air is forced. The aerobic environment supports a different, but also common microorganism that, like the anaerobic process, feeds on the organic fraction of the waste. The waste is converted to by-products that include CO_2, water vapor, and compost. Typically, a site had to be located in a rural area; otherwise, the odors from the process could become a nuisance.

Biochemical

The biochemical process is based on breaking down the cellulosic part of the organic fraction of the waste stream. This would include certain foods (e.g., vegetables, fruits), paper products, and yard vegetation. Biosolids can also be added as a waste material. All other materials in the waste stream should be removed prior to the process.

In the process, following drying and shredding of the waste, the prepared waste stream is mixed with water and sulfuric acid in a closed reactor vessel. This causes a reaction that in conjunction with common bacteria already in the waste breaks down the material into sugar compounds and a by-product known as lignin. There are some companies that are testing natural enzymes, instead of the strong acid chemical, to initiate this reaction.

The resulting sugar compounds and water are sent to a fermentation unit where yeast is added. The yeast reacts with the sugars to convert them to alcohol. The alcohol mixture is then heated and distilled to remove the solids. The resulting distilled alcohol (grain alcohol or ethanol) can be used as fuel. The lignin by-product is sent to a gasifier where it is used to produce heat for the drying process or can potentially be further processed for use as a fuel substitute in power plants [17]. A basic biochemical process is shown in Figure 4.39.

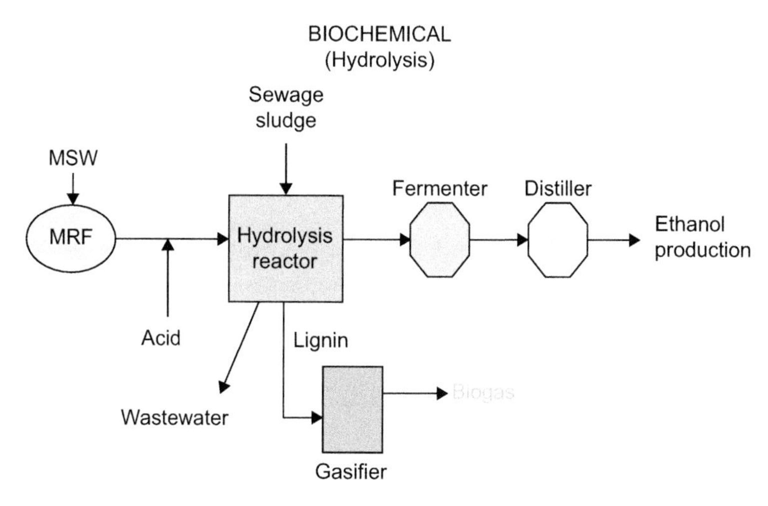

Figure 4.39 Basic biochemical process. (From SCS Engineers)

Status of Commercial Operating WC Facilities

Plasma Arc Gasification

As shown in Table 4.10, there are four operating plants utilizing MSW as feedstock. Only one of these, the Utashinai City plant, can be considered commercial; the others have been only operated as pilots or intermediately operated for testing purposes. A pilot plant in Ottawa, Canada is currently being tested by Plasco Energy and has only been intermediately operated with a maximum continuous runtime of 36 h using a presorted, postconsumer waste stream as feedstock. Plasco is currently in the process of converting this plant to commercial operations, having successfully negotiated an operating contract with the City of Ottawa.

Pyrolysis Plants

As given in Table 4.11, the use of pyrolysis technologies to process MSW has occurred mainly in Japan and Germany where these plants reportedly process about two million tons of materials per year. There are no commercially operated facilities in the United States, although a pilot facility was operated in Green Bay, WI using American Combustion Technology pyrolytic systems for testing purposes. Oneida Seven Generations, Inc. has plans to construct a pyrolysis facility using 148 Tpd of MSW and 61 Tpd of plastic waste in Green Bay, WI.

There were a number of full-scale MSW pyrolysis demonstration plants, which were constructed in the United States during the late 1970s and early 1980s by Monsanto and Union Carbide. These facilities were not commercially successful and were eventually shut down. Similarly, a 91 Tpd MSW pyrolysis facility was constructed in New South Wales, Australia in 2001 by Brightstar Environmental [20]. This facility incorporated the use of an autoclave process where the organic fraction

Table 4.10 Commercial Operating Plasma Arc Gasification Facilities

Location	Throughput (Tpd)	Owner/ Operator	Technology Supplier	Start of Operation	Feedstock
Yoshi, Japan	25	Hitachi Metals, Ltd.	Westinghouse Plasma Corp.	1999	MSW
Utashinai City, Japan	200	Hitachi Metals, Ltd.	Westinghouse Plasma Corp.	2003	MSW
Mihami-Mikata, Japan	22	Hitachi Metals, Ltd.	Westinghouse Plasma Corp.	2002	MSW Biosolids
Ottawa, Canada	94	Plasco Energy	Plasco Energy	2007	Shredded MSW Shredded plastics

Source: From Ref. [11].

Table 4.11 Commercial Operating Pyrolysis Facilities Using MSW

Location	Throughput (Tpd)	Technology Supplier	Start of Operation
Toyohashi City, Japan	440 77 (bulky waste)	Mitsui Babcock	2002
Hamm, Germany	353	Techtrade	2002
Koga Seibu, Japan	286 (MSW and biosolids)	Mitsui Babcock	2003
Yame Seibu, Japan	242 55 (Bulky waste)	Mitsui Babcock	2002
Izumo, Japan	70,000 TPY	Thidde/Hitachi	2003
Nishiburi, Japan	210 63 (bulky waste)	Mitsui Babcock	2003
Kokubu, Japan	178	Takuma	2003
Kyouhoku, Japan	176	Mitsui Babcock	2003
Ebetsu City, Japan	154 38 (bulky waste)	Mitsui Babcock	2002
Oshima, Japan	132	Takuma	2003
Burgau, Germany	154	Techtrade	1987
Itoigawa, Japan	25,000 Tpy	Thidde/Hitachi	2002

Source: From Refs. [11,19].

was dried before being sent to a pyrolysis vessel. This facility operated for only 6 months and was shut down due to its failure to meet permitted conditions.

Anaerobic Digesters

There are nearly 240 AD facilities around the world with operating capacities greater than 2500 Tpy. These plants process not only the organic fraction of the MSW waste stream but also organic waste from food industries and animal manure. Europe

Table 4.12 European Countries with AD Facilities

Country	No. of Plants	Country Capacity (Tpy)
Germany	55	1,250,000
Spain	23	1,800,000
Switzerland	13	130,000
France	6	400,000
Netherlands	5	300,000
Belgium	5	200,000
Italy	5	160,000
Austria	4	70,000
Sweden	3	35,000
Portugal	3	100,000
United Kingdom	2	100,000
Denmark	2	40,000
Poland	1	20,000
Total	127	4,605,000

Source: From Ref. [21].

leads in the number of AD plants and total installed capacity principally due to the European Union Directive that requires member states to reduce the amount of landfilled organics by 65% by 2020. As shown in Table 4.12, there are more than 120 plants processing the organic fraction of MSW in Europe of about 4.6 million Tpy. The principal technologies used around the world are provided by the following companies: Dranco, Kompogas, Linde, RosRoca, Valorga, BTA, and Cites.

Currently, there is only one commercially operated AD facility in the United States, which is located on the campus of the University of Wisconsin-Oshkosh. It processes about 6000 tons of yard and food wastes per year. Further, an AD facility digesting source-separated organics has been commercially operating in Toronto, Canada for a number of years processing about 90,000Tpy. A second AD facility is currently under construction in Toronto and should be operating within a year. Similar AD facilities have been authorized by Monterey Solid Waste Authority, the City of San Jose, Quebec City, and Montreal, with additional facilities funded in the Province of Quebec.

Benchmark Metrics

Minimum Waste Throughput Processing Capacity

Waste Conversion Based on author's experience, typical WC technologies, including thermal, biological, and biochemical are represented to operate on a comparatively smaller scale. Between these types, generally, thermal technologies require more significant waste throughput to be economically viable. Due to lesser equipment and energy requirements, biological technologies can generally support smaller waste throughput.

Thermal By a wide margin, the greatest amount of recent activity in WC technology is with the thermal technologies, dominated by the plasma arc conversion process. This is mainly due to its potential for large power production and overall reduced air emissions.

The lack of an operational track record for both large-scale and small-scale WC technologies suggests that a WC technology plant should more likely be planned initially as a small pilot plant. A pilot plant, based on proven laboratory and mini-pilot scale technology, would be no more than about 100Tpd with the potential for scale-up should the technology be proven at the pilot stage and with regional collaboration.

Anaerobic Facilities In Europe, the anaerobic process has been used successfully to process MSW. The sizes of these plants reportedly range from 3000 to 182,000Tpy. Converted to a daily capacity, and assuming a 6-day/week processing schedule, these capacities range from 10 to 580Tpd.

As noted in the paragraphs above, there are many operating AD facilities in the size range potentially generated by many communities. These facilities are successfully processing from about 15 to 250Tpd of food and yard wastes diverted from residences, restaurants, and businesses and converted into methane that is used to produce power.

Summary of Readiness for Commercial Operations

Some, but not all of the alternative WC technologies are ready for commercial operation. Table 4.13 summarizes the technologies discussed herein and whether, in the author's opinion, they are ready for pilot plant or commercial operation on a scale necessary to serve most towns or regions.

As depicted, each of the seven technologies have demonstrated pilot plant readiness either nationally or internationally; however, only three of the technologies appear ready for commercial scale operations. These three technologies are the biological processes and the autoclave process. With the exception of the autoclave, each of these technologies requires preprocessing requirements to remove potential contaminants from the incoming waste stream.

Table 4.13 Summary of Main Processes

Process	Preprocessing	By-Product	Primary Product	Pilot Plant Readiness	Commercial Readiness
Pyrolysis	High	Ash	Syngas/oil	Yes	No
Gasification	Medium	Ash/slag	Syngas/char	Yes	No
Autoclave	Low	None/recyclables	Pulp	Yes	Yes
Anaerobic digestion	Medium/high	Filtrate water	Biogas/ compost	Yes	Yes
Hydrolysis	High	Waste water/ash	Ethanol	Yes	No
Aerobic digestion	Medium/high	None	Compost	Yes	Yes
Plasma gasification	Claims low/high	Slag	Syngas	Yes	No

Capital and Operating Costs

As described, due to the relatively recent development of the alternative WC technologies, there are few, if any, full-scale operational plants in the United States. Thus, there are no reliable figures readily available for capital and operating costs.

Two large, comprehensive studies were conducted as part of a detailed review of alternative WC technologies in the United States [22,23]. The on-going studies were sponsored by Los Angeles County California, as continuation of that region's program initiated in 2003 to further address the regions acute problems with energy pricing and availability, air quality, traffic congestion, and reliance on landfills that had limited useful life. The original study screened 27 technologies in the initial phase (2005) and reduced the list to five "finalists" technologies in the subsequent 2007 report. The finalists are currently planning to build small-scale demonstration plants to prove their respective technologies.

Although there have been other large alternative technology screening/evaluation studies conducted, the L.A. County studies seem to have the most detailed information on projected U.S.-based plant costs and economics. Table 4.14 summarizes the project economics for five finalist biological and thermal alternative WC technologies that were developed as part of the L.A. County study in 2007.

The costs and economic summaries were provided by the selected technology vendors, using some pricing assumptions for specific items provided by the planning committee and applicable to southern California only. The consultant retained by L.A. County conducted an independent review of the costs and economics provided by the vendors and concluded that the figures provided were, in general, reasonable estimates that matched with the independent assessment's conclusions.

Tipping Fee Survey

Table 4.15 compiles costs from the 2005 and previously discussed 2007 L.A. County studies. The middle column are tipping fees summarized from the economic projections rendered in the 2005 study, which had similar pricing and cost assumptions as in the 2007 follow-on study. Tipping fees in the 2005 study ranged from $61 to $197 per ton for the eight vendors. Two plants exhibited tipping fees in the $50 to $70 per ton range, while six were higher than that.
L.A. County considered a tipping fee in the range of $50–$70 per ton, to be competitive with the tipping fees charged by the large regional landfills serving the area. Table 4.15 indicates that two of the four thermal technologies and one anaerobic technology provided costs that indicated the plant could offer a tipping fee in the $50–$70 per ton range.
The difference in tipping fees from 2005 to 2007 probably reflects some differences in the pricing assumptions in individual studies including: proposed plant capacities were larger in 2007 and purchase pricing structure for the power produced was revised. It is also assumed that the market conditions for the development of these plants from 2005 to 2007 likely became more favorable as basic energy costs in the United States continued escalating.

Comparison to WTE Fees

Because conventional WTE plant technology has been in existence for decades, with hundreds of plants operating in the United States and abroad, comparative cost

Table 4.14 Summary of Project Economics for Thermal and Biological Conversion Technologies

Technology	[a]Annual Throughput (TPY)	Projected Design Capacity (TPD)	Capital Cost ($)	Capital Cost per Ton ($/ton)	[b]O&M Costs ($)	Total Costs ($)	Estimated Annual Revenue ($)	Estimated Net Costs ($)	Calculated Tipping Fee ($/ton)	Tipping Fee Variation ($/ton)	Adjusted Tipping Fee ($/ton)
Biological (anaerobic)	100,000	300	21,000,000	70,000	4,900,000	8,170,000	3,000,000	5,170,000	52	6	58
Thermal (autoclave)	51,100	200	35,000,000	175,000	9,000,000	13,100,000	8,400,000	4,700,000	92	0	92
Thermal (pyrolysis gasification)	80,000	242	30,140,000	125,000	5,580,000	7,740,000	3,280,000	4,460,000	56	2	58
Thermal (gasification)	97,000	312	75,200,000	241,000	11,000,000	20,700,000	8,000,000	12,700,000	131	1	132
Thermal (gasification)	138,000	413	56,600,000	137,000	8,260,000	14,200,000	6,300,000	7,900,000	57	12	69

Source: From Ref. [23]
[a]Tons per year (Tpy), demonstration plant only.
[b]First year costs only, does not include annual debt service.

Table 4.15 Summary of Economic Data[a]

Technology	[a]Projected Design Capacity (TPD)	[a]Calculated Tipping Fee ($/ton)	[b]Calculated Tipping Fee ($/ton)
Biological (anaerobic)	100	93	58
Biological (anaerobic)	100	67	–
Biological (anaerobic)	100	197	–
Thermal (autoclave)	–	–	92
Thermal (plasma arc)	100	172	–
Thermal (gasification)	150	61	58
Thermal (gasification)	300	186	132
Thermal (pyrolysis–gasification)	100	129	69

Source: From Ref. [23]
[a]Tons per year (Tpy), demonstration plant only.
[b]Adjusted tipping fee from Exhibit 12–16, based on phase II study.

information is more established; although a completely new WTE plant has not been constructed in the United States in more than 10 years. Figure 4.40 shows a visual summary comparison of the tipping fees estimated for the alternative WC technologies in the L.A. County studies and the tipping fees for operating WTE plants. The graph shows that the appropriate "average" tipping fee for a WTE plant is about $60 per ton. The estimated "low" and "high" range is estimated to be from about $35 to $80 per ton, respectively.

The tipping fee ranges for alternative technologies are provided as a crude comparison to the WTE tipping fee. A large tipping fee range, from low to high, is evident. These plots reflect expected uncertainties and risks at the time of the studies, which would not be unusual for technology that is still in the development or pilot plant stages. Most WTE plants in the United States have a capacity anywhere from 500 to about 4000 Tpd, and this affords them a valuable "economy of scale" over the much smaller proposed alternative technologies.

Such a large range of tipping fees for alternative WC technologies may not actually be the case if a study were done today. Projected tipping fees are a function of many regional cost factors, including:

- Power production/quality and quantity of syngas
- Air emissions and treatment
- Market for by-products
- Downtime/equipment reliability
- Preprocessing requirements (sorting equipment, MRF, etc.)
- Operator experience
- Financial contributions by vendor
- Contractual obligations
- Ancillary costs (transmission line, etc.).

However, in the author's opinion, these summary costs for alternative WC technologies suggest that tipping fee ranges are likely to be somewhat higher than a

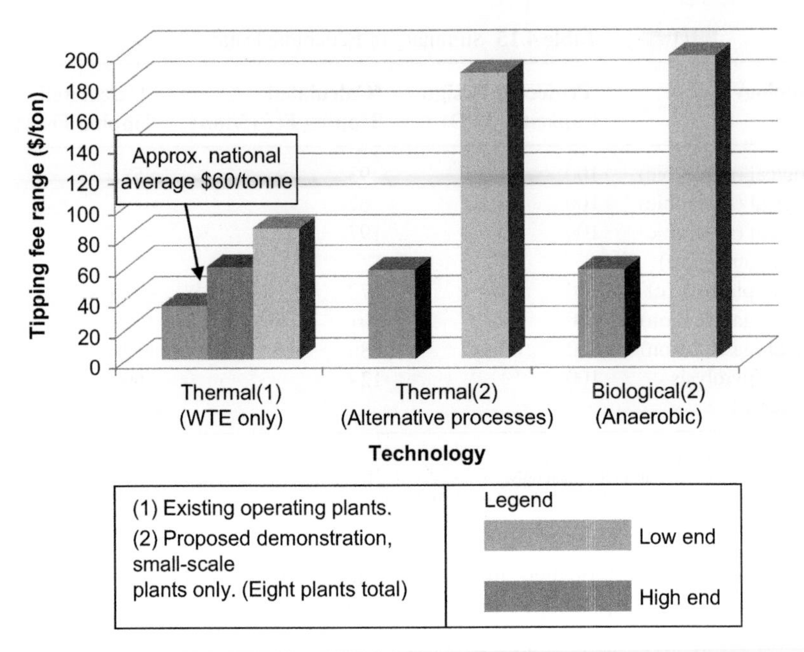

Figure 4.40 Summary of tipping fee range for technologies.

WTE plant, until enough of the plants are operating and hard costs are generated to validate that they can operate at a tipping fee comparable to a WTE plant.

Potential Fuel Products

The greatest opportunities for market growth for alternative WC technologies appear to be producing advanced ethanol products. The U.S. EPA has promulgated the Renewable Fuel Standard (RFS), which mandated that all gasoline in the United States is blended with 10% ethanol to produce E10 gasoline. At the time of this writing, this is the highest blend level approved for use on all American passenger cars and light trucks. There is some concern by automobile manufacturers and the public with the potential corrosive effects of increased ethanol in the fuel mix. This has limited the expansion of the market for E15 and E85 (15% and 85%, respectively) ethanol blend.

Nonetheless, in November 2012, the U.S. EPA reiterated is commitment to the biofuels industry by denying a requested waiver of targets set out in the RFS (Table 4.16). This has sent a strong message of support to potential producers of these advanced ethanol products. Further, demand for these fuel products could see substantial increases in demand as the U.S. Department of Defense procures more renewable fuels.

Advantages and Disadvantages

All of the alternative WC technologies have some potential benefits and disadvantages. The over-riding aspect of all of the alternative WC technologies is that they

Table 4.16 RFS Biofuels Mandate (Billions of Gallons)

Year	Total Renewable Fuels	Cap on Corn Starch-Derived Ethanol	Portion to Be from Advanced Biofuels			
			Total Non-Corn Starch	*Cellulosic*	*Biodiesel*	*Other*
2006						
2007						
2008	9.00	9.00	0.00	0.00	0.00	0.00
2009	11.10	10.50	0.60	0.00	0.00	0.00
2010	12.95	12.60	0.95	0.01	1.15	0.20
2011	13.95	13.20	1.35	0.01	0.80	0.30
2012	15.20	13.80	2.00	0.01	1.00	0.50
2013	16.55	14.40	2.75	1.00	1.28	0.75
2014	18.15	15.00	3.75	1.75		1.00
2015	20.50	15.00	5.50	3.00		1.50
2016	22.25	15.00	7.25	4.25		2.00
2017	24.00	15.00	9.00	5.50		2.50
2018	26.00	15.00	11.00	7.00		3.00
2019	28.00	15.00	13.00	8.50		3.50
2020	30.00	15.00	15.00	10.50		3.50
2021	33.00	15.00	15.00	13.50		3.50
2022	36.00	15.00	21.00	16.00		4.00

Source: From Ref. [24].

are relatively new and thus do not have a "track record" from which one can derive hard conclusions related to actual, proven benefits and disadvantages. So we can only postulate what the actual advantages, disadvantages, and economics might be. This exercise is based on assessing the information available from vendors, review of operational history for some very small-scale pilot plant facilities that may have operated intermittently, and evaluation of these technologies that are processing waste streams other than a normal mixed MSW.

Table 4.17 summarizes the advantages and disadvantages of the alternative technologies and the WTE technology. We offer the following generalized conclusions, in addition to the comments in the table, about the viability of the technologies:

- *Biological (anaerobic)*. Commercial scale proven at smaller capacities (i.e., 200–300 Tpd) in Europe. Developing a consistent market for the compost by-product is a major challenge and affects the operating economics. Only a few small-scale plants are currently planned in the United States.
- *Thermal*. Generally unproven at a commercial scale. One small pilot facility (85 Tpd) is operating in Canada. A complex process that must be optimized to provide the desired high-quality synfuel. There is much planning activity in the industry, and in the next 5 years, there will likely be some operational plants to better demonstrate the potential scalability and viability of these technologies.

Table 4.17 Advantages and Disadvantages to Waste Processing Technologies

Process	Advantages	Disadvantages
Thermal—pyrolysis/ gasification	Potential for high power production, high conversion	Untested, possibly high O&M costs, ash disposal
Thermal—autoclave	Provide higher quality recyclables	Lack of market for compost
Biological—aerobic	Proven, "low" tech. Emissions less of a concern.	Some odor; Lack of market for compost, low conversion
Biological—anaerobic	Low emissions, low odor	Lack of market for compost
Plasma gasification	Potential for high power production, high conversion	Untested, possibly high O&M costs, safety concerns, slag market (?)
Biochemical (hydrolysis)	Fuel production, biosolids processing	Untested, treats only cellulosic part of waste
WTE plant	Proven large-scale technology	Large volumes of unusable ash, costly air emission control systems

- *Biochemical.* Unproven at a commercial scale. A few plants have been planned but have been delayed. Tied to the dynamic market for ethanol and competition with many other processes that do not use MSW.

References

[1] SCS Engineers. Solid waste and recycling plan. Municipality of Skagway, AK: prepared for the Municipality of Skagway; Skagway, Alaska 2013.

[2] Berenyi EB. Materials recycling and processing in the United States: 2007–2008. Westport, CT: Governmental Advisory Associates, Inc; 2007.

[3] Egosi N, David W. The MRF of tomorrow. Resource Recycling. March 2010; p. 13–17.

[4] Edmonds R. The state of the news media 2012: an annual report on American journalism. Pew Research Center's Project for Excellence in Journalism, <http://stateofthemedia.org/2012/newspapers-building-digital-revenues-proves-painfully-slow/newspapers-by-the-numbers/>. Accessed June 2, 2013.

[5] U.S. Department of Commerce. U.S. Census Bureau News: quarterly retail e-commerce sales. Ist Quarter 2013, <http://www.census.gov/retail/mrts/www/data/pdf/ec_current.pdf>. Accessed March 6, 2013.

[6] Kessler Consulting, Inc. Materials recovery facility technology review. Prepared for Pinellas County Department of Solid Waste Operations. September, 2009. Clearwater, Florida.

[7] de Thomas D. Sorting it out. Resource Recovery. April 2012; p. 28–32.

[8] SCS Engineers. Colombia integrated waste NAMA feasibility study for Colomba-Guabal Landfill, Cali, Colombia and Los Pocitos Landfill, Barranquilla, Colombia. Prepared for the Center for Clean Air Policy. May 2013. Washington, D.C.

[9] SCS Engineers Comprehensive review of solid waste collection and disposal options. North Carolina: Town of Chapel Hill; 2012.
[10] Yepsen R. BioCycle. January 1, 2012(53).
[11] SWANA Research Foundation. Waste conversion technologies. December 2011. Silver Spring, MD.
[12] Griffith-Onnen I, Zak P, Jennifer W. On-site systems for processing food waste. Northeastern University, Report prepared for the Massachusetts Department of Environmental Protection. April 2013. Boston, Massachusetts.
[13] McGinley M, McGinley C. Measuring composting odors for decision making. Presented at U.S. Composting Council, 2005 annual conference. San Antonio, TX: January 25, 2005.
[14] California Integrated Waste Management Board. Conversion technologies status update survey. April 2009. Sacramento, California.
[15] Federal of Canadian Municipalities. Solid waste as a resource: guide for sustainable communities. 2012. Quebec, Canada.
[16] New York City Department of Sanitation. Evaluation of new and emerging solid waste management technologies. September 2004. New York, New York.
[17] Affiliated Engineers. Alternative energy analysis, the University of North Carolina. July 2010. Raleigh, North Carolina.
[18] Youngs H. Waste-to-energy in California: technology, issues, and context. California Council on Science and Technology. October 2011. Sacramento, California.
[19] U.S. Department of Energy. Draft environmental assessment for Oneida Seven Generation Corporation, energy recovery project, Green Bay, WI. 2011. Washington, D.C.
[20] Global Alliance for Incinerator Alternatives. Incinerators in disguise. April 2006.
[21] Levis JW, Barlaz MA, Themelis NJ, Ulloa P. Assessment of the state of food waste treatment in the U.S. and Canada. Waste Manage 2010;30(8–9):1486–94.
[22] Los Angeles County. California, California conversion technology evaluation report. Phase I assessment. October 2005.
[23] Los Angeles County. California, California conversion technology evaluation report. Phase II assessment. October 2007.
[24] Renewable Waste Intelligence. Waste to biofuels market analysis 2013. December 2012.
[25] U.S. Energy Policy Act of 2005, 63040. Federal Register November 7, 2005;72(215), <www.epa.gov/epp/pubs/guidance/fr72no215.pdf>. Accessed April 10, 2013.

5 Marketing Recyclables

Market Forces

Before discussing specific materials and their markets, it is useful to consider some basic issues which affect the consumption (i.e., availability of end-user markets) of recycled materials. On the whole, the US manufacturing sector has been geared to the use of virgin materials. Product manufacturers, in response to society's consumer attitude, have long been more attuned to consumer convenience (especially since the end of World War II) than to product durability and reusability. However, in recent years, as waste disposal costs continue to escalate and the public is becoming more aware of the beneficial impacts of recycling, more manufacturers are addressing recycling and reuse in a positive manner. Still, though, there are significant barriers which apply to the reuse of reclaimed materials. These limitations can be broadly grouped into three categories:

1. Limitations affecting the demand of materials in industry
2. Limitations affecting the supply of materials to industry
3. Limitations imposed by government.

These limitations are interrelated and greatly affect the supply and demand of the recyclable materials market.

Limitations Affecting the Demand of Recyclable Materials in Industry

Figure 5.1 shows the demand curve for recyclable materials utilized by industry to produce final products. This curve demonstrates how much a particular recyclable material industry is willing to purchase at a given price. The shape of the demand curve (nearly vertical) demonstrates that the demand for recyclable materials is relatively inelastic (i.e., does not change), with respect to supply. Demand for recyclable materials is dependent largely upon production capacity. Other factors affecting industries' demand for recyclable materials are manufacturing capital costs, potential contamination problems, and other costs associated with using reclaimed materials.

Manufacturing Capital Costs

Many manufacturers have complete "in-house" manufacturing processes from virgin materials procurement to final product shipment and may be unwilling to alter these processes in order to accept recyclable materials, for which they have less control

Solid Waste Recycling and Processing. DOI: http://dx.doi.org/10.1016/B978-1-4557-3192-3.00005-1

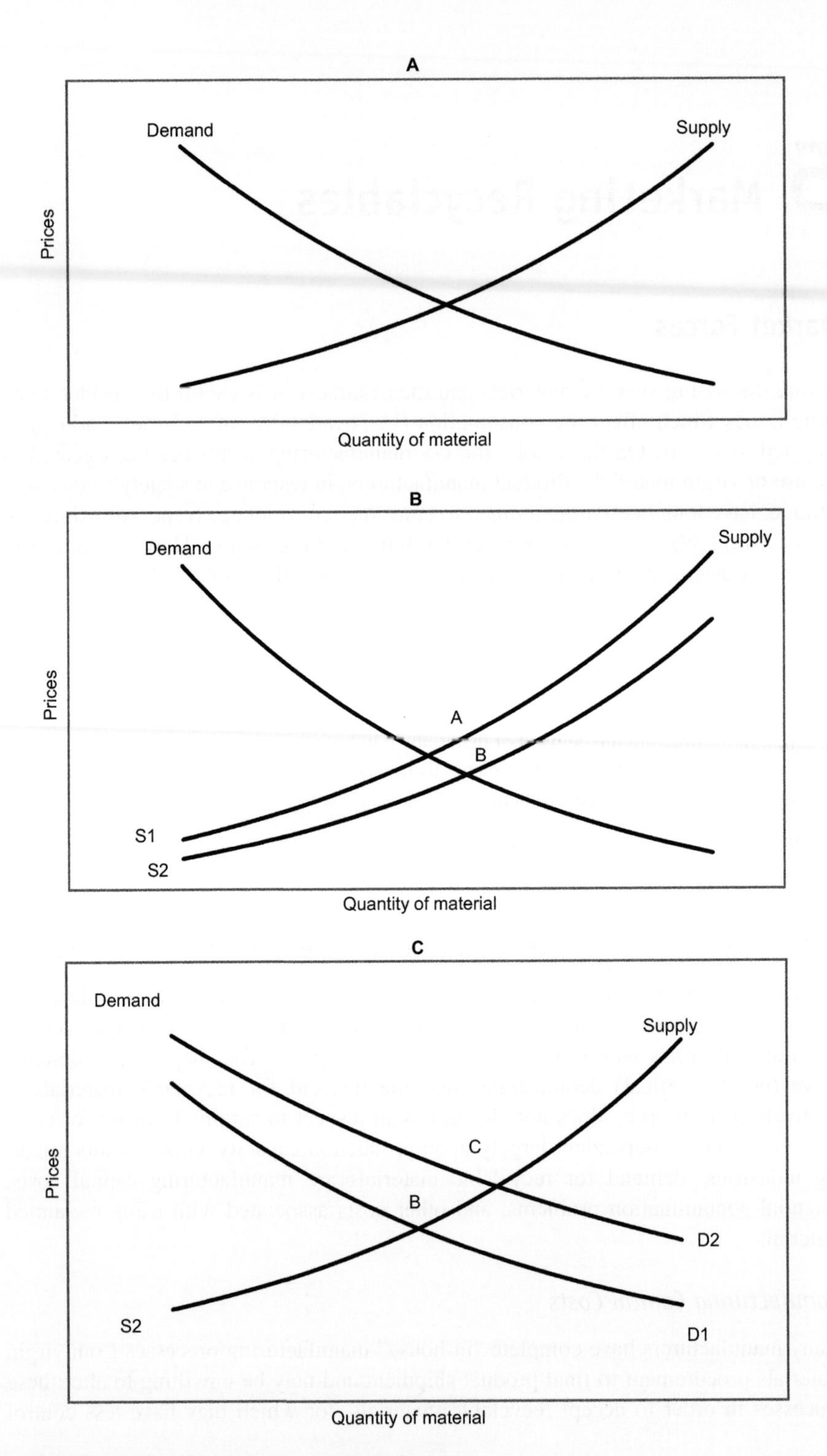

Figure 5.1 Supply and demand curves for recyclables.

over quality, quantity, and availability. The expense of current technology has prevented many industries from retrofitting their existing manufacturing process, currently using virgin materials, to processes which utilize reclaimed raw materials. High capital costs required to expand production capacity in order to accept more recyclable materials has also limited the demand.

Potential for Contamination

Raw material specifications can often be met by both virgin and recycled materials; however, the buyer, wary of potential contamination problems, may impose stricter standards before purchasing recycled materials. Large quantities of virgin raw materials that meet material specifications can be procured from a single source, whereas reclaimed raw materials must first be collected from a large number of sources, processed, and distributed to a few manufacturers. As the number of sources increases, so does the possibility of contamination.

Other Costs

Other costs associated with the use of reclaimed raw materials are increased inspection costs which are accrued during every step of the manufacturing process from collection to final product approval. As a result of the increased production costs associated with reclaimed materials, many of the products produced are more expensive than those produced from virgin materials. In today's economically driven society, there is seldom a large demand for an equal, but more expensive product. However, many recent public opinion polls indicate that the public is becoming more aware of the beneficial impacts of recycling and would prefer to buy products packaged in containers made from recycled products.

Limitations Affecting the Supply of Recyclable Materials to Industry

Figure 5.1B shows the supply curve for recyclable materials delivered to manufacturers. This curve indicates the quantity of recyclable material that suppliers are willing to supply at every possible price during a given period of time. The supply of recycled materials is shown to be extremely elastic because it is relatively easy and inexpensive to get into the supply side of the business. These elastic and inelastic characteristics of supply and demand combine to cause frequent and often volatile shifts in material prices and supply quantities over short time periods. The recent decline in the price of plastics in the United States due to changes in Chinese port inspection practices is a good example. Also, waste streams vary seasonally. It follows, then, that the availability of certain recycled materials also fluctuates. For example, more beverages are consumed in summer than in winter. This results in a higher amount of glass and aluminum beverage containers available for capture and reuse during the summer. This type of fluctuation may make recyclable materials less attractive to an industry that is looking for a stable, consistent supply of material.

Another limitation faced by recycled material suppliers are transportation costs. Natural resources occur in concentrated form, whereas recycled secondary materials from waste are dispersed and have high-attendant collection costs.

Limitations Imposed by Government

Up to this point in the discussion, the recovering and reuse of recyclable materials could be thought of as a simple supply and demand interaction, where the supply of recyclable materials delivered to manufacturers equal the demand for those recyclable materials. Point "A" on Figure 5.1B represents a typical equilibrium point. As recently as 5 years ago, many recyclable markets were felt to be at or near their equilibrium points.

Since that time, however, a great deal of national, state, and provincial legislation has been enacted which establishes recycling goals for recyclable materials. To meet these newly mandated recycling goals, communities across the country have established recycling and public awareness programs to promote recycling and reuse. The result of these programs was a dramatic increase in the amount (i.e., supply) of recyclable materials available to an industry which did not (and still does not) have the production capacity (demand) to utilize the materials in many cases. This set of events is shown graphically in Figure 5.1C. Here, the original supply curve S_1, has shifted right to S_2, and the equilibrium point moved from point "A" to point "B." The new equilibrium point B shows a slight increase in the quantity of materials reclaimed and a dramatic drop in price paid for those materials.

To counteract this ominous trend, many state legislatures are beginning to enact legislation to build a recycling infrastructure to help stimulate the market for the purchase of recycled materials. In many states, this legislation has involved development of a multifaceted strategy which has included grant programs, investment tax credits, loan guarantees, start-up and expansion loans, changes in permit programs for the siting of recycling facilities, information and market clearinghouses, technical assistance programs including workshops and market studies, and changes in state and local procurement requirements which encourage buying recycled products. Another major market impact by government can be the imposition of overall state recycling goals and specific product recycling levels. The intent of all these steps taken by the government is to provide the economic incentive to industry to expand its production capacity by shifting the demand curve for postconsumer recycled products to the right. Thus, increasing material supplies resulting from recycling collection programs will have viable markets as recyclable materials' prices will be driven slowly upward.

Recyclable Materials

Waste Paper

Waste paper, a significant portion of the solid waste stream, is bought and sold on the basis of grade, and prices vary accordingly. Grades of paper range from low, such as newspaper and corrugated, to high, such as printing, writing, and computer paper.

Mixing different grades lowers the quality by reducing the value in remanufacturing. Paper grades are generally defined as specified by the Paper Stock Institute of America which lists specific guidelines that define different grades of paper based on type and preparation. In general, the source of the secondary fiber will dictate the paper grade into which it can be processed.

Paper and paperboard products are distinguished by the physical properties they possess. Generally, the physical properties of paper products are dependent upon two primary factors: fiber length and pulping method. Fiber length determines characteristics such as strength, stiffness, opacity, and printability. Basically, as the ratio of long fibers to short fibers increases, so does the strength; however, this increase in strength is also accompanied by a reduction in surface smoothness. This becomes a disadvantage from a printing standpoint because ink prints more uniformly on a smooth surface. For this reason, packaging materials (e.g., corrugated) are composed of long fibers, while printing and writing papers are composed of short fibers.

The pulping process controls characteristics such as cleanliness and brightness (whiteness). There are three types of pulping processes: mechanical, chemical, and a combination of the two. Pulp produced from mechanical pulping (groundwood pulp) produces a weak paper and usually requires the addition of some chemical to hold the paper together as it travels through printing presses. Groundwood pulp contains lignins which provide high bulk and opacity as well as smoothness and ink absorbency, which makes groundwood pulp ideal in the manufacturing of newspapers. Incidentally, it is the lignin that turns yellow when a newspaper is exposed to the sunlight.

Chemical pulping removes the lignin during a cooking process, and as a result, produces a strong pulp that can be bleached to high levels of brightness. Bleached chemical pulp, known as kraft pulp, is used in the production of fine-white writing paper. Unbleached chemical pulps are used in the production of brown paper bags and corrugated boxes.

Old Newspapers

Old newspapers (ONPs) are a major fiber resource for the paper industry, both domestic and abroad, and are bought and sold as a commodity. The United States consumes about 10 million metric tons of newsprint a year or about 64 pounds per person per year with about 60% supplied from Canada [1]. Newspaper is still one of the most visible forms of waste papers found in the household and is, by both weight and volume, one of the highest percentages (generally 10%) of most residential waste streams. Commercial establishments also generate large volumes of newspaper, but the percentage in commercial waste streams is much lower than that in residential waste streams, and is not as easily recovered.

To a household consumer, newspapers consist of newspaper, supplements, magazines, and retail store tabloids. To the end users of recycled newspaper and domestics and foreign paper mills, newspaper has a different meaning. They consider ONP as a raw material and the many inserts as contaminants. Whether the material is wet or dry and baled or unbaled are also criteria by which they judge ONP. When baled, ONP should contain less than 5% "other paper." Prohibitive materials may not exceed 0.5% and outthrows may not exceed 2%. Prohibitive materials are contaminants, such as glass, plastic, metal, or other nonpaper material. "Outthrows" refer to types of paper other than the one being specified. These guidelines show that recycling ONP requires a careful effort to produce quality reusable material.

Other paper products such as magazines, telephone books, glossy inserts, and junk mail present several other problems. During the recycling process, ONP is reduced to a pulp by adding warm water and deinking chemicals. This pulp is then washed and screened to remove any rags, glass, plastics, and dirt. Newspaper normally loses between 15% and 20% of its weight due to the clay-coated paper on which they are printed. The clay coating is separated from the paper during the process and is washed away as waste. The combined loss could be 30–40% of the weight of original fiber. The latex adhesives in magazines and telephone book bindings create latex balls in the repulping process which cannot be removed by screening. They can cause spots on the finished product or may cause the paper to stick together in a paper roll.

ONP is recycled into many products. It can be recycled into newsprint and paperboard which is used in production of game boards, book covers, boxes, photo albums, tubes for toilet paper and paper towels, and construction materials such as roofing felt paper, insulation, and wallboard. ONP has also been used for products like cellulose insulation, animal bedding etc.; however, those markets are small regional markets and are not expected to increase in size.

ONP is the main grade of waste paper collected from households. During the 1970s and early 1980s, scout troops, schools, religious, civic charitable organizations, and a few private individuals collected ONPs and sold them to raise funds for local projects and programs.

In the mid-1980s, increased demand by a few recycled newsprint mills and an expanding export market caused an increase in the volume and price of recovered ONP. The volume of recovered ONP continued to increase from 4.3 million tons in 1986 to 7.37 million tons in 2010, roughly a 72% recovery rate [2].

The weak economy has not only reduced the number of newspaper subscribers, but also has decreased the size of most newspapers as a result of reduced advertisement. Newspaper readership in the United States has declined dramatically since 1990 with page counts on most newspapers plunging almost 47% in the recent decade [2]. Most industry observers suggest that the impact of electronic media has had a dramatic negative impact on the tonnage of newspapers generated and recycled in the last decade [2].

To increase the usage of ONP in recycled products, state governments have established mandatory recycled content requirements for many paper products or entering into voluntary agreements with newspaper publishers. Some 27 states have voluntary or mandatory recycled fiber requirements for newspapers. Further, many local, state, and federal agencies, as well as many private firms, are establishing self-imposed purchasing requirements, which require a certain percentage of their paper products to be made of recyclable material.

Over the next 5 years, it has been projected by the paper trade associations that ONP capacity will grow insignificantly as new or retrofitted mills come on line outside the United States, primarily in China. Exports in the United States are almost half of the market for ONP [3].

Old Corrugated Cardboard

Old corrugated containers, referred to as OCC, represent the largest single category of waste paper collected for recycling. In the United States, OCC comprises over 40% of all waste paper recycled, and in some large metropolitan areas, over 60%.

To understand OCC, the word "corrugated" means "to form or shape into wrinkles or folds or into alternative ridges or grooves." This definition explains the center portion of OCC or what is known as the fluting or corrugated medium which forms a continuous pattern of grooves. The fluting is placed between two sheets of paperboard to provide strength to a container. This material is commonly called cardboard.

Corrugated containers are a major source of waste in a municipal waste stream. Most OCC is generated by commercial establishments such as supermarkets, restaurants, department stores, and various other retail stores. Many large stores keep the OCC separate and sell it to paper dealers or bale it themselves and sell it directly to the mills. The residential segment generates only a small portion of the OCC in the total waste stream, and most of it is contaminated after coming in contact with the other mixed wastes.

Corrugated, like all waste paper used in recycling, must be kept dry and free of contaminants. Both of these requirements must be met if a recycling effort is to realize the maximum value of the OCC. Moisture is a factor that affects the weight of the corrugated cardboard. Prices paid for corrugated are generally by the ton; therefore, OCC must be dry for an accurate weight.

Because most of the corrugated generated for recycling comes from retail establishments, common contaminants are plastics such as trays used to package food items; Styrofoam packing materials; plastic bags, wrap, film, and cups; metal objects such as wire hangers, case strappings, can, and nails; plastic and wax-coated cartons such as those commonly used to pack fresh produce (distinguished by a very dark brown color and shiny surface); and other contaminants, including floor sweepings, wood, food waste, beverage cans, and trash. Contamination is a serious concern of the recycling industry, since unwanted materials adversely affect production efficiency and finished quality.

Corrugated is used primarily in production of paperboard products for packaging where the strength of boxes, cartons, and fiber cans is very important. Paperboard production is divided into four main categories: (1) unbleached kraft used for outer facing of corrugated and solid fiber boxes, (2) semichemical used primarily for center fluting in corrugated boxes, (3) solid bleached paperboard which is converted into packages such as milk cartons, frozen food cartons, and containers for moist, liquid, and oily foods, and (4) recycled paperboard used in folding cartons and set-up or rigid boxes.

Unbleached kraft paperboard has traditionally been a virgin fiber product that contained very little waste paper. However, waste paper use in unbleached kraft mills has increased in recent years from 2% in 1970 to about 9% of the approximately 31.9 million tons produced in 2011 [1]. OCC and corrugated box plant clippings are the predominantly used recycled fibers in paperboard production. Continued research into the use of corrugated in unbleached kraft paperboard indicates that this sector of the industry will account for a significant increase in the use of waste paper in paper products.

Semichemical paperboard has traditionally required the use of corrugated waste paper in the range of about 20%, virtually all of which has been from box plant cuttings. The demand for corrugated waste paper in semichemical paperboard had increased to 40% of total fiber in 2010 [4]. This is a much higher ratio than most experts expected and is due to the installation of cleaning equipment for processing postconsumer waste paper. The use of OCC in semichemical paperboard is expected to increase slightly in the future as more mills add cleaning equipment.

Bleached paperboard is solid white throughout and made from primary fiber with little or no waste paper used in its manufacture. Since the largest market for bleached paperboard is in packaging liquid, moist, or oily food, sanitary requirements are of prime importance. Waste paper in bleached paperboard production is likely to come only from clippings and cuttings of paper and paperboard that have been handled, treated, and stored in a clean, sanitary manner. Recycled paperboard traditionally has been the only paper product made from 100% waste paper.

Since 1960, OCC generation has skyrocked in the United States, increasing by 296%, thereby increasing its share of the municipal solid waste (MSW) waste stream by 27%, according to the US EPA [3]. However, being easily recyclable, the recovery rate has increased by 150% over the past 50 years. According to the American Paper Institute (API), the recovery rate for OCC was nearly constant at 91.2% for 2011 [4].

China, Japan, and Korea are among the largest importers of US waste paper which they use to make corrugated containers to ship TVs, stereos, VCRs, etc., to foreign markets, including America. Further, demand for box production in the United States has rebounded from the Great Recession, benefitting in from the need for smaller shipping containers by online merchants such as Amazon, and Barnes and Noble.

High-Grade Paper

High-grade paper is a generic name of a wide variety of office papers such as writing, computer, and copying paper. Often known as office white paper (OWP), it is made of long, high-quality fibers that are brighter than newspaper and packaging grades. This grade can be further divided into two categories: pulp substitutes and deinking high grades. Pulp substitutes are clippings and shavings from items such as envelopes, bleached paperboard cuttings, business forms, ledgers, and other high-quality fibers generally derived from computer paper centers, print ships, and other paper converting plants. These types of high-grade paper can be used as a direct pulp substitute for virgin material.

Deinking high grades are usually paper of the same high-quality fiber as pulp substitutes, but they have gone through a printing process. They are generally derived from printing plants and office buildings that use high-grade paper such as ledger or other forms of high fiber bleached paper. This paper must first go through a deinking process before it can be used as a pulp substitute. Although high-grade waste paper represents a small portion of the total export tonnage, it is in greater demand because of its high quality and, therefore, commands a higher price per ton.

High-grade clippings and cuttings as well as deinking high grades are usually reprocessed back into their original forms. For example, white ledger clippings are repulped and used to make new white ledger paper. Another common use is in tissue paper.

The value and recyclability of high-grade paper depends on its being separated by grades based on fiber length and on its being free of all unwanted material. The key to recovery and sorting high-grade paper lies in producing mill-specific paper bales. Contaminants are dependent upon how the mill plans to reclaim the fibers. Different cleaning and/or deinking systems are used to produce different paper products. As a result, what is considered a contaminant at one mill may not be considered a contaminant at another mill. Various contaminants include rubber bands, yellow post-it tabs, laser, fax, and photocopy papers, newspaper, magazines, and cardboard. Paper clips and staples can be magnetically removed during the shredding process.

OWP has two distinct markets, which makes analyzing future markets extremely difficult. They are low-quality OWP and high-quality OWP. The low-quality OWP is used in the production of paperboard products and packaging materials. These products are quite often a mixture of high grades and other waste paper. The use of the low-quality OWP helps to maintain the quality of the end product. The high-quality OWP is used for printing and writing paper, tissue, and market pulp. The remainder of this section discusses high-quality OWP since low-quality OWP can be introduced into many existing processing facilities.

In the past, high-grade paper markets were established for preconsumer white and colored ledger although a small amount of postconsumer scrap was also bought and sold. Prices for the different types of high-grade waste paper vary greatly and range from $30 to $200 per ton.

Over the past two decades, many voluntary office recycling programs have been established. The result of these collection programs has been the supply of OWP increased by 3.7 million tons since 1990 in the United States, while at the same time office paper generation has declined by 2.2 million tons or 29% in the last decade as the use of electronic files and e-mail has spread through most office environments [4].

Residential Mixed Paper

Residential mixed paper (RMP) is a paper category which covers a wide range of waste paper types and is not limited to grade or fiber content. RMP is composed of magazines, catalogs, direct mail, and boxboard packaging (e.g., cereal and shoe boxes). RMP is generally recovered from garbage, which makes it difficult to separate the contaminants. Collection of RMP would require a great deal of public education and can be very costly.

Mills that use RMP usually like to know the content of the mixed paper and generally deal with processors who consistently provide them with good mixed paper that will meet their specifications. The paper stock standards and practices circular PS-86 defines mixed paper as follows: "a mixture of various qualities of paper not limited to type of packing or fiber content. Prohibitive outthrows may not exceed 10%." Metal, glass, plastic, and other nonpaper material are considered as contaminants. Mixed paper is used almost exclusively in recycled paperboard and construction grades, which include products such as roofing felts, door coverings, automotive fibers, and pipe coverings.

Markets for RMP in the United States and overseas are primarily the mills that produce recycled paperboard and construction grades. Exported waste paper is consumed primarily by countries that do not have abundant supplies of virgin fiber materials but which have growing economies that have created a demand for paper and paperboard products. The largest importers of US waste paper are Mexico, South Korea, Taiwan, Japan, Italy, and Venezuela. In many of these countries, recycling rates for waste paper are very high; because virgin fiber is not added, the quality of the recycled paper and paperboard decreases each time it is used. As an alternative, imported high-quality waste paper is used to help maintain the quality of the finished product.

Because the United States has a good supply of high-quality, inexpensive waste paper, the export market has continued to grow. Some export markets like India, with low labor costs have found it feasible to buy mixed paper and then separate it manually into specific grades for other uses. In the past, RMP played an important role when demand for recycled paperboard products was high. However, as source-separated materials such as newspaper and corrugated become increasingly available to the waste paper market, the use of mixed paper began to decline. Mixed paper value has always been less than source-separated materials like newspaper, office paper, and corrugated.

The future for RMP looks bleak as long as there remains a glut of source-separated materials in the market. However, the future for magazines will increase as newsprint facilities which utilize new flotation deinking technology come online. This new deinking technology requires 20–50% coated magazine stock to aid in deinking and production of a stronger sheet. When the market price for corrugated increases, corrugated medium manufacturers and recycled boxboard manufacturers are likely to become interested in purchasing small quantities (10–20%) of the less expensive RMP.

Glass Containers

Glass, which has been in use for thousands of years, is a transparent substance, made primarily from sand, soda ash, and limestone. Glass containers are produced in three colors: clear (flint), brown (amber), and green. Of these colors, flint has the largest number of applications and is usually in greatest demand by glass manufacturers. Brown or green glass is used in products where exposure to sunlight may cause the product to degrade.

The primary glass produced in the waste stream is the glass container, which is mainly composed of soda bottles, beer bottles, and condiment jars. Other glass products such as cooking ware, dishware, ceramics, windows, and specialty glass also appear in the solid waste stream but are considered as contaminants due to their chemical composition or heat-resistant properties.

Most manufacturing facilities involved in glass recycling use only bottles and jars, i.e., container glass. These manufacturers also require collected glass to be separated by color, since the material is used to make glass of the same color. Mixing colors produce a low-quality glass container and in many cases an esthetically unappealing end product.

If a manufacturer does not have appropriate processing equipment, the recycler (or middle processor) is required to remove metal, paper, and other glass contaminants from the container glass as well as separate them according to color. Since glass furnaces operate at temperatures of 2600°F, most metals will melt and corrode the furnace linings. Other metals such as aluminum form small balls that end up in finished products making them unusable. Melting mixed colors of glass and glass of varying chemical compositions in the same batch can lead to a foaming action in the furnace which produces off-color bottles with numerous air pockets within the glass. Ceramics and heat-resistant glass do not melt at the temperatures used in a glass container furnace and show up in the end product as "stones" or other defects.

The primary markets for recyclable glass containers are the 75 glass container manufacturing plants in the United States. Other secondary markets include road construction, either on the surface called "glassphalt" or as a roadbase aggregate; filler aggregate in storm drain and French drain systems; the fiberglass industry; glass beads for reflective paints; abrasives; foam glass and other building materials.

In 1967, 40 container glass manufacturers produced glass from 112 plants in 27 states. Today 17 companies operate 54 facilities in 27 states. As glass containers lost market shares to aluminum cans, PET, and other plastic materials over the last two decades, the glass container industry has consolidated and reduced capacity. Tonnage has declined by 2.5 million tons since 1990 [5].

Three companies (Owens-Brockway, Gallo, and Saint-Gobain), now supply about 90% of glass container demand (9.36 million tons or 60.6 pounds per person per year) in the United States in 2010, estimated at approximately 25 million glass containers, with nearly 75% beer bottles; the remaining mostly food containers [5].

In recent years, factors that have contributed to the increase in glass recycling are preservation of natural resources, reduction in litter, energy conservation reduced waste quantities, disposal cost, and reduction of raw material use. The natural resources in glass manufacturing are sand, limestone, and soda ash. Although these resources are abundant in the United States, they are geographically separated by long distances, which leads to high transportation costs in procuring these raw materials. Thus, using recycled glass helps conserve oil and gas. The "bottle bill" legislation passed by many states in the 1970s encouraged glass manufacturers to use reclaimed ground glass called "cullet." Using cullet allows furnaces to operate at lower temperatures which extend furnace life, reduce energy costs, and lower stack emissions. The use of cullet in the manufacture of glass has increased steadily from 22% in 1988 to 33.4% in 2012 [5].

Other considerations influencing private and government involvement in glass recycling are limited landfill space and increasing costs of waste disposal. By promoting recycling, the glass industry is also better able to maintain its share of the container market which is being impacted by the aluminum and plastic container industries.

The price paid for glass containers is determined by color, quality, and the extent to which it has been prepared (i.e., crushed or whole). The prices paid for glass containers vary greatly depending upon proximity to glass manufacturing facilities. In some locations, some collectors are paying $5–8 per ton to market their green glass. The unstable market for green glass cullet is, in large part, due to the manufacturers. The quantity of imported foreign liquors bottled in green glass exceeds the production capacity for green color containers.

Probably the greatest influence on cullet and bottle prices in the late 1980s was the supply of new material from communities with mandatory recycling programs. As more communities implemented recycling programs to extend the life of existing landfills, a new flood of glass caused prices to decrease. Because mandatory recycling was motivated by cost avoidance, communities were willing to give glass away, if necessary.

For decades, intermediate and secondary processors produced cullet from natural resources and reclaimed materials, which was then sold to glass manufacturing facilities. However, in the 1980s, due to the abundance of available container glass, many glass manufacturing plants spent millions of dollars on glass benefication facilities at their plants. The introduction of new buyers has helped to stabilize the market prices.

With cullet prices heading downward in most parts of the country, many communities are contemplating removing glass containers from their recycling programs. Communities have discovered that glass recycling is a labor-intensive and time-consuming endeavor which can prove to be expensive with falling cullet prices. In many cases, however, the public relations benefits and avoided tipping fees are felt to outweigh the collection and processing costs.

The future market value for recyclable container glass is dependent on three issues: new processing strategies, developing new markets, and recycled content legislation. Manufacturing companies are currently experimenting with increased levels of reclaimed cullet and utilizing mixed color cullet in batches to ease the saturated markets.

Scrap Metals

Two types of metals are commonly recycled: ferrous metals and nonferrous metals. The nonferrous metals include materials such as aluminum, brass, copper, lead, and zinc. By far, the most common recyclable of this group is the aluminum beverage can, which is discussed separately in the next section. Ferrous metals include cast iron, "tin cans," and other steel products and stainless steel, including industrial scrap, car bodies, and household appliances.

Scrapped autos are a major source of ferrous scrap for the secondary metals market. On an average, an automobile remains on the road for about 10 years. This time is increasing because of special anticorrosive treatment of metals and greater precision and longevity of the power train. Studies done in the early 1980s of all deregistered cars indicate that 60% are processed by scrap shredders and then sold as ferrous scrap to iron and steel mills. Another portion of deregistered autos is stocked by the auto dismantling industry as an inventory of used parts.

Since ferrous scrap from junked autos and other industrial scrap represent a significant portion of ferrous that is already shipped to the steel industry, this section focuses on another major source of ferrous scrap, the ferrous can. In the residential waste stream, the tin-plated steel food can is the largest volume ferrous metal product discarded.

Ferrous cans are made up of three general types of containers. The predominant type, in terms of quantity, is the tin-plated steel food container that is coated on the inside with a thin film of tin to preserve freshness of food. The second type is the steel can that does not have the tin plating. The third type is the bimetal can, used primarily for beverages, that has a steel bottom and sides and an aluminum top with a pull tab.

The steel industry prefers the tin-plated steel container after the tin and other contaminants are removed in a detinning facility. In some cases, the industry will accept small quantities of either tin-plated or bimetal cans directly from scrap dealers or municipal recycling programs. These containers should be baled or densified. Detinners (companies that remove the tin) prefer the tin cans free of labels, thoroughly rinsed out, and shredded in 1–2 in^2 of thin plate. In some cases, the containers can simply be flattened and shipped loose with labels attached.

The only domestic source of tin is in the millions of tons of postconsumer tin-plated steel food cans, which has given rise to the detinning industry. After the tin is removed by dipping the container into a chemical solution that separates the tin from the steel, it is reused for its original purpose, tin plating. About 6 pounds of tin is recovered from a ton of scrap cans. It is important to note that the detinning industry serves the purpose of upgrading the quality of the steel by removing tin and other contaminants. The resultant steel is a fairly high and uniform grade that can be used in electric arc furnaces (minimills) or the traditional basic oven furnace to produce any type of steel desired. Some tin-plated steel is purchased directly from suppliers without the tin removed, but it can only be used in very small quantities.

The detinning industry views the increased interest in recycled tin cans as a potential source of additional tin. An industry representative indicated that there was an opportunity for a fivefold increase. As more recycling programs include this material, additional detinning plants may go online. The potential exists for continued and expanded recycling of tin-plated steel cans.

Bimetal cans pose a far more difficult problem; until the industry finds an economical way to separate aluminum from steel, the potential recycling of these containers is very limited. This limitation may be offset by the current trend of reduced demand by the beverage industry in using this type of container because it is not easy to recycle. This has also resulted in the sudden shift to the all-aluminum can in certain markets.

The backbone of the source-separated tin can recycling market was the detinning industry. However, the number of detinning plants fell from 15 in 1968 to seven in 1987 as a result of competition in the beverage container market from plastic and aluminum. Sufficient capacity exists in the detinning industry to accept additional tin cans. However, it is difficult to economically collect and process tin cans. As the cost of solid waste disposal increases, solid waste officials are becoming more interested in removing tin cans from the waste stream.

In the past, the steel industry preferred to buy the higher grade steel from the detinning industry after the tin was removed. Tin is a contaminant and gives steel a flaky or cobbled appearance. However, recently steelmakers are taking note of the scrap metal can market for a variety of reasons. Steelmakers' efforts to regain a larger share of the beer and soft-drink can markets include more recycling of bimetal beverage containers (steel body with aluminum lid). Purchase of these scrap cans demonstrates the industry's action to affect this recycling as well as reduce litter and lessen the requirements for landfilling. Steel mills have overcome such problems as the need to minimize residual tin. The steel can, among the lowest cost source of ferrous scrap, is a desirable scrap if it is relatively free of contaminants and if its content is limited properly in the scrap mix. Lastly, a growing amount of scrap is available from source separation operations and other sources, such as waste-to-energy plants that take in MSW and magnetically separate steel cans and other ferrous materials for resale to scrap markets.

Aluminum

Recycled aluminum has long been the highest-valued commodity of all the secondary materials. This is due largely to the high cost of mining and processing virgin material, the large investments in capital equipment, and high amounts of energy needed to smelt primary aluminum. With the high costs in processing aluminum and demand for all-aluminum cans growing, recycling has helped aluminum manufacturers remain competitive with other forms of beverage packaging. The 95% reduction in energy consumption to recycle aluminum cans coupled with a highly competitive foreign bauxite market makes recycling the most viable alternative for the domestic aluminum industry.

Bauxite, the principal raw material from which aluminum is made, is one of the most abundant components of the earth's crust, and the technology to mine it has been known for approximately 100 years.

Domestically, bauxite deposits are very limited with the US aluminum industry depending on foreign sources for most of its supply. The high costs of mining and processing aluminum ores caused the industry to realize that scrap aluminum had value as a resource and that it could be collected and remelted for use in new products. The technology for recycling scrap aluminum was developed about 75 years ago, but did not include recycling cans, which is a relatively new development brought about primarily by trends in the beverage packaging industry.

Over 200 registered aluminum alloys, each having different mixtures of components, are broadly classified as wrought and cast aluminum. Wrought aluminum (e.g., beverage cans) is used primarily to produce beverage containers and foils and is collected for reuse in like products. Cast aluminum consists of scrap castings that are reprocessed into ingots of casting alloy. Both components of aluminum in the waste stream are reviewed below for their recyclability.

Aluminum Cans

The aluminum can, commonly called UBC (used beverage can) in the aluminum industry, continues to dominate the beverage can packaging market with an average share of over 95%. UBCs are often collected, baled, and shipped to processors; however, two of the largest recyclers of aluminum beverage cans, ALCOA and Reynolds Aluminum, prefer to receive their UBC loose and flattened in tractor trailers so they can examine them for contamination. Contamination is anything that is not an aluminum beverage can: dirt and moisture on cans, plastic, steel, iron, lead, paper, or other common items. In cases where large quantities of aluminum will be recovered, processors often provide the communities or organizations with the necessary processing equipment (i.e., flattener, blower, and tractor trailers). Recycling plants may return highly contaminated shipments at the shipper's expense.

UBCs are currently collected with other types of aluminum (small quantities of aluminum foils are acceptable) and are used mainly in the manufacturing of new aluminum cans. Once collected, the UBCs are shipped to conversion mills to be remelted and formed into high-quality aluminum sheet metal. These sheets are then sold to can manufacturers to produce new aluminum beverage containers. Recent recycling rates for UBCs (65%) indicate that a single can may be recycled as many as six times in 1 year. The aluminum industry has encouraged and promoted recycling by developing aluminum buyback centers and providing UBC processing equipment and transportation to companies, groups, and individuals interested in recycling.

When the all-aluminum can was first introduced into the packaging market in 1963, it capitalized on the knowledge that recycling aluminum only takes 5% of the energy than producing virgin material from bauxite. Since the bauxite supply in the United States was and still remains very limited, recycling the aluminum can made it possible for domestic aluminum companies to decrease dependence on foreign bauxite supplies and to help conserve our own natural resources. In the 1970s, many states started antilitter campaigns and passed legislation requiring consumers to pay deposits on beverage containers. This increased the recovery rate of UBCs and helped to make recycling of UBCs possible.

Historically, the UBC market has been dominated by the large aluminum processors such as Alcan, ALCOA, and Reynolds which have been able to minimize the competition for UBCs from secondary smelters and export markets. However, in the past few decades, these previously excluded industries have begun to purchase UBCs from smaller sources such as small communities and individuals.

The price for UBC can varies from day to day, based on the price for primary aluminum, the quality of the collected UBC, and the supply and demand of UBC on the market. Due to these factors, it is impossible to get accurate past prices for the UBC.

The same factors that influenced prices in the past will continue to have control on future pricing, but the implementation of municipal, county, and statewide recycling will also continue to have a dramatic effect. In the past, UBC recycling was done predominantly by the private sector and financed by the aluminum recycling industry. However, mandatory recycling, unlike private recycling, is based on preservation of landfill space and cost avoidance associated with landfill costs. With the increasing supply of UBC on the market and fewer profit-motivated parties involved, the aluminum recycling industry may be able to reduce prices while receiving increasing supplies of UBC. This, however, will not cause an oversupply of aluminum cans. Theoretically, the aluminum industry could recycle every aluminum can made provided it is prepared properly.

Other Aluminum

In addition to UBCs, MSW also includes other aluminum products (e.g., aluminum foil, lawn furniture, window frames, outdoor play sets, and automobile parts). Except for UBCs and aluminum foil (which are wrought aluminum), the above-mentioned products are cast aluminum. The difference between wrought and cast aluminum is primarily how it is manufactured into finished products. Wrought products, like aluminum cans and foil, are stretched and pressed by a mechanical process to form the product. Cast aluminum is melted and poured into forms or casts and cooled to harden into shape.

Since the can recycling industry deals primarily with UBC, the secondary aluminum sectors deal with many of the remaining 200 or more alloys. Aluminum by itself does not provide the strength needed for the uses stated above. Other metals are therefore blended in to form an aluminum alloy. The chemical composition of each alloy is different to provide the specific characteristics needed for each product. For instance, aluminum is mixed with iron and manganese and is used in automotive parts to help reduce weight in cars, while providing needed strength and rigidity.

Aluminum foil is almost pure aluminum. Since it is free of most alloying metals, processors primarily use it as a "sweetener" in their scrap furnaces. The influx of pure aluminum into the scrap aluminum batches "dilutes" the metal alloys and "aids" in the cleansing process.

Industries and households sell their aluminum to a scrap dealer, who generally separates it into different grades for sale to the secondary aluminum industry. Sorting is usually done by scrap dealers who know the industry and the type of alloys. They buy most kinds of ferrous and nonferrous metals, separate and upgrade them, and then sell them to their prospective markets. Unlike many recyclables that go through little processing before being returned as finished products, secondary aluminum requires as many as four processes before it is completely recycled.

A large number of products are produced from the various alloys of aluminum. The secondary aluminum industry produces different grades of casting and extrusion ingots that are sold to foundries and die casters for uses in many products. One of the largest buyers of secondary aluminum is the auto industry. Transmissions, engine blocks, brake components, pistons, and other automotive parts are all cast from aluminum or aluminum alloys. The demand for specific grades of secondary aluminum is determined by the properties offered by various alloy chemical compositions.

Demand for recycled or raw scrap aluminum is growing in foreign countries such as Japan, Korea, and Taiwan which are known for their automobile production.

The value of remelting secondary aluminum into ingots for new products was realized in the early 1900s by the secondary aluminum industry. This industry was the primary market for scrap aluminum until the last 20 years. In the late 1960s, demand for secondary aluminum alloys grew due to the increased uses of aluminum in automobiles. Demand continued into the 1970s, and new competition entered the secondary market from primary producers adding equipment that used secondary scrap as well. Foundries and fabricators also added scrap aluminum capacity.

Primary aluminum prices have historically influenced aluminum scrap prices on a day-to-day basis and will continue to play a major role in the very volatile pricing of aluminum. The recent weak dollar overseas has stimulated countries like Japan to import US secondary aluminum for its automotive needs.

The main factor affecting other aluminum scrap prices is the United States and world economies. The United States and many European countries have had several recessions in the last two decades. These recessions influence the monies people have, or are willing, to spend on items such as homes and automobiles which are large markets for the aluminum industry. Other products such as lawn furniture and outdoor play sets have also seen reduced sales. Automotive design trends imposed by federally mandated mileage standards have increased the use of aluminum per passenger car by some 30% over the past two decades.

Plastics

Plastic resins are synthetic materials made from oil and natural gas that are combined in a polymerizing process. Each resin has a different molecular structure that gives the material unique qualities and its value as a material. Today, more than 200 different types of resins are used to produce plastic products within the United States. This large number of resins had hindered the plastic recycling effort due to identification problems, until the Society of Plastics announced its "voluntary coding system" (which appeared on the bottom of many plastic containers) in 1979. Plastic packaging materials such as containers, films, and wrappers provide several benefits which have broadened their use. These include breakage and leakage resistance, hot/cold insulation properties, versatility in shapes and sizes, and reusability (e.g., using grocery bags or bottles for other purposes). However, the durability which has given plastic its advantage is also a disadvantage. When disposed, much of this packaging is not biodegradable and persists in the environment for a long time.

The primary types of plastic resins used in containers include the following:

PET	Polyethylene terephthalate
HDPE	High-density polyethylene
PVC	Polyvinyl chloride
PP	Polypropylene
LDPE	Low-density polyethylene

These five plastic resins compose roughly 98% of the plastic containers manufactured worldwide today. Plastic containers tend to be made from these readily identifiable resins and represent approximately 49% of the plastic packaging. Film plastic, such as plastic bags, accounts or 37%, while coatings and closures are 9% and 5%, respectively.

A 2012 survey conducted by the American Chemistry Council (3903 cities and counties representing about 80% of the US population) indicated that at least 94% of the US population have access to PET and HDPE bottle and cap recycling and just over 57% have access to all plastic bottles and all nonbottle rigid containers. A majority of cities and counties continue to use the resin identification codes to describe acceptable materials for recycling [6].

PET

PET is the clear plastic resin used to make beverage bottles and food containers. It was first patented in 1973 and first recycled 4 years later. PET is a strong but lightweight form of polyester used for soft-drink bottles, liquor bottles, and other food and nonfood containers. Recycled PET is used to make soft-drink bottles, other containers, fiberfill in jackets and sleeping bags, carpet fibers, industrial strapping, and many other consumer items.

In 2011, plastic soft-drink bottles were recycled at rates exceeding 29% in the United States and resulted in 780,000 tons of PET and 127 million pounds of HDPE from the base caps. However, market research studies conducted by the plastics industry suggest that a potential market existed for 700 million pounds of PET from recycled soft-drink bottles in nonfood applications [6]. Thus, in the opinion of the plastics industry, the market is far from being saturated.

HDPE

There are two types of HDPE bottle grade material: the homopolymer and the copolymer. The homopolymer HDPE (blow molded) has a stiffer molecular structure and is used for dairy, water, and juice bottles. The copolymer HDPE (injection molded) has a more flexible molecular structure and is chemically more resistant to bottle contents such as detergents and household cleaners. These two types of HDPE are incompatible and cannot be mixed together.

PVC

PVC is known for its flexibility, high chemical resistance, and lower cost. It is mainly used in flexible bags and piping materials; however, it has also been used for food jars and bottles. PVC has limited thermal stability and tends to degrade quickly when processed. These three types of resins represent nearly 94% of the plastics market in the United States [6]. HDPE and PET are included in most recycling programs, while PVC is usually excluded due to its low market share.

The best markets for plastics currently are for separated plastic compounds. Therefore, easily identifiable and separated plastics are most readily marketable.

PET soda bottles and HDPE milk, juice, and water bottles are in this category and are the most commonly recycled postconsumer plastics.

The fact that plastic containers are a high-volume, low-weight material affects the costs of collection from the community, processing, and shipment to market. Collected plastics require manual sorting by trained people, by material type and by color. Many plastic processing centers are hesitant to purchase postconsumer PET from programs where sorting was carried out buy nontrained personnel for fear of PVC contamination. It takes only a few PVC bottles to contaminate a 40,000 pound truckload of PET.

Baling the sorted material is one approach to reducing the volume. A more effective approach is grinding the material in a granulator. This, in particular, increases the importance of enforcing strict separation procedures. A careful assessment of the volume of plastic containers generated and collected from a community will influence processing, along with location of the buyer. The primary buyers of plastic waste from municipalities are few, and volume reduction through baling or grinding is essential to reduce transportation costs.

Since World War II, plastics have been used in an increasing number of products due to their versatility, durability, and lightweight characteristics. The introduction of the plastic (PET) soda bottle caused a phenomenal amount of growth within the plastic packaging industry.

The history of the plastics recycling industry has been primarily propelled and shaped by three main factors: deposit and redemption mechanisms, mandatory state recycling goals, and the public relations factor due to the recyclability of aluminum and glass beverage containers. In the early to mid-1980s, several states enacted "bottle bills" which first focused on postconsumer plastic recycling. The material returned from these programs was extremely contaminant free. This facilitated the marketing of the material, quantity recovered by these programs was very low which increased the expense of transportation.

Recent state-mandated recycling goals have led to numerous recycling programs across the nation. These programs have produced a glut of reusable PET and HDPE plastics available to the plastic resin markets. Some communities are fortunate enough to live in proximity to plastic resin producers thereby receiving some monetary value for their collected plastics. Many are forced to dispose of their collected plastics in landfills because it is less expensive than paying the transportation costs.

For the past several years, recycled PET has been used in the manufacture of new bottles. Before March 1991, these bottles were limited to nonfood and nonbeverage applications (fiberfill, strapping, carpeting, engineering plastics, and geotextiles). Thus, while no federal law specifically restricts the use of recycled plastic in food and beverage packaging, packagers and resin producers concluded that cleaned postconsumer scrap converted into pellets and mixed with virgin materials cannot be assured to be contamination free. At this same time, market pressures were being asserted on soft-drink giants to get a recycled content bottle on the store shelves next to the recycled aluminum, steel, and glass containers. As a result, chemical recycling or regeneration was born, and shortly thereafter Coke and Pepsi began using recycled content bottles.

Polypropylene

Recovered PP has been next bid low-hanging fruit for plastics recycling [7]. In recent years, PP has been grabbing market share in food packaging as new packaging technologies enable manufacturers to change from metal cans to PP cans. PP is usually the resin of choice for yogurt, ketchup, and mayonnaise bottles, and flip-top lids from tubes of toothpaste and suntan lotion. For example, one-third of the soup cans from Campbell Soup Company are currently PP containers. Easily recoverable PP rigid plastics are often found in backrooms of deli, bakery, and seafood departments of medium to large supermarkets in the United States. These include frosting pails, salad bar containers, fish containers, and pharmaceutical stock bottles. These currently represent the best opportunity of large volumes of PP that are very clean, natural or white, and mostly food-grade quality [8].

According to data released by the American Chemistry Council, the amount of nonbottle PP rigid plastics has nearly tripled from 129 million pounds in 2009 to 363 million pounds in 2010 [9]. The most extensive corporate recycling effort in the United States is being done by Preserve Products (Waltham, Massachusetts), which makes kitchen, personal care, tableware, and food storage containers from 100% PP. Its "Preserve Gimme 5" recycling program, which has been done in partnership with Whole Foods Stonyfield, and Brita, has grown from 65 Whole Food supermarkets in 2009 to 245 Whole Food stores in 35 states. This program has teamed with Recyclebank to offer curbside recycling participants points and rewards for PP plastic products recycled through Recyclebank.

Regeneration is the process of breaking down plastics into its two monomers (dimethyl terephthalate, DMT, and ethylene glycol, EG) by adding chemicals and heating. The DMT is then purified to produce near virgin quality PET resin. Regeneration processes are costly requiring tremendous amounts of energy and very clean scrap plastics. Regenerated PET bottle resin has been estimated to cost 10–50% more than a virgin PET bottle. This cost factor appears to be the reason why no one is yet producing a new bottle with greater than 25% regenerated resin. Regeneration has many critics stating that plastics regeneration is merely a public relations effort by the soft-drink industry to introduce a recycled content bottle to the public.

Regeneration accepts only extremely clean scrap and is too costly to produce bottles with more than 25% regenerated resin. Regeneration of plastics might actually hinder future recycling efforts by keeping market prices at low levels. These low prices may discourage cities from collecting more plastics until they know there is an established market. At the same time, processors are reluctant to invest in equipment without an assured flow of quality materials. Further, another threat to development of this market is the refillable PET bottle which can be cleaned and refilled up to 25 times before being recycled. This new container is now used in several counties and may be introduced in the United States in the next few years.

Notwithstanding these problems, plastic manufacturers have come under increasing pressure in recent years to use the recovered materials from community-based plastic recycling programs. The industry has funded an extensive research program nationally to study collection and sorting technology at the Center for Plastics Recycling Research at Rutgers University —The State University of New Jersey. As an outgrowth of this research at Rutgers, a number of resin reclamation technologies have been developed and offered for commercialization.

The production of products from recycled commingled plastic materials is a growing industry. Developing markets for first-ever products to replace wood, concrete, and steel with plastics has been an inherently slow process, but has been stimulated with new specialized extruder/molder technology developed in recent years. Several manufacturers have reported plans for increased production capacity to produce products such as plastic lumber, picnic tables, landscaping timbers, fencing, pallets, car stops, and park benches.

Lastly, any discussion on plastics would not be complete without any mention on recent bans on several plastic waste streams such as Styrofoam and plastic bags.

Styrofoam

Styrofoam is a trademark brand of extruded polystyrene foam, which is manufactured by Dow Chemical. The product was originally designed for use in thermal insulation. The process of production has been slightly modified for production of coffee cups, restaurant “take-out” containers, cushioning material for use in packaging (“peanuts”), and insulation in coolers. Due to littering across the landscape and its low degradation, environmentalists have promoted bans on this material in the United States and Canada. A number of North American cities have already enacted Styrofoam bans to prohibit retail sales of foam products such as plates, cups, bowls, peanuts, and coolers. Recycling of this material is only in its infancy with many community recycling coordinators promoting instead reuse of Styrofoam for the following projects:

- Craft projects
- Floaters for fishing
- Drainage in potted plants
- Creating sets for theatrical productions
- Model train communities
- Stuffing bean bags or bean bag chairs
- Packaging for shipping.

Plastic Bag and Film

Recovered plastic film enters the market typically in a variety of different categories and HDPE, LDPE, and LLDPE resins:

- Commercial film (clear)—Clear, clean PE film including stretch wrap and poly bags.
- Commercial film (mixed)—Mixed color PE film with no postconsumer bags.
- Mixed film—Mixed color, clean PE film including retail collected postconsumer bags, sacks, and wraps.

Table 5.1 Postconsumer Film (Pounds) Acquired for Recycling in the United States and Canada

Year	Exported	Purchased for Use in the United States and Canada	Total
2011	426,738,000	583,023,000	1,009,761,000
2010	455,984,000	515,823,000	971,807,000
2009	490,718,000	363,659,000	854,377,000
2008	469,968,000	362,426,000	832,394,000
2007	462,611,000	367,569,000	830,180,000
2006	221,082,000	590,928,000	812,010,000
2005	183,701,000	468,776,000	652,477,000

Source: From Ref. [9].

- Curbside film—Mixed PE film generated at materials recovery facilities (MRFs).
- Dirty agriculture film—From the ground with up to 50% contamination.
- Clean agriculture film—Dry and from uses that do not touch the ground with up to 10% contamination.

"Stretch" film is oftentimes collected alone or mixed with other polyethylene film, including postconsumer bags, sacks, and wraps.

Table 5.1 gives a breakdown of postconsumer films recycled in the United States and Canada based on surveys conducted by Moore Recycling Associates Inc., for the American Chemistry Council. These data also illustrate a continued trend toward more domestic processing of plastic films over the past 7 years; an estimated 58% in 2011. Composite lumber manufacturers were reported to consume the majority of domestic postconsumer film and bag material. The remainder was exported, primarily to China for reprocessing into a variety of products.

These surveys suggest that most film reclaimers try to keep their processing costs low and avoid washing in favor of further sorting. Nearly 90% of the market is currently focused on fairly high-quality film so they can bypass the wash phase. As of now, the composite lumber industry appears to be the leading US market for postconsumer film.

Plastic shopping bags have become a common sight at grocery stores, pharmacies, and convenience centers. However, there has been a significant pushback from the general public and environmentalists about the impact of discarded plastic bags on the environment and overall public litter. As a result, many communities have enacted bans on plastic bags. Overall, more than 60 California cities have enacted plastic bag bans. Where these bans have gone into effect, the reusable bag has exploded with many businesses putting their logo on reusable bags [10].

Carpet

Postconsumer carpet is commonly manufactured into engineered plastic resins or used in the production of new carpet fiber or carpet backing. The Carpet America

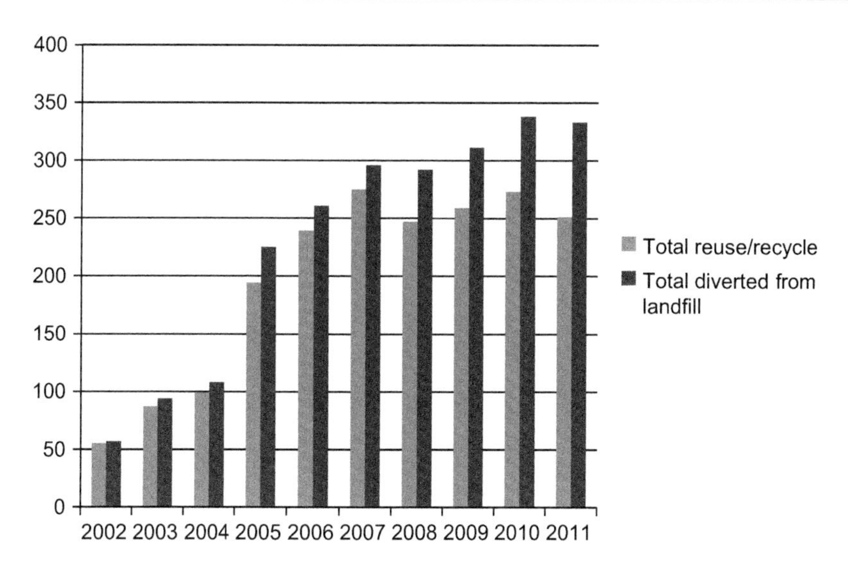

Figure 5.2 Postconsumer carpet reuse and recycling/diversion in the United States, 2002–2011 [11].

Recovery Effort (C.A.R.E.) serves as the carpet industry's professional trade association. Its mission is to encourage design for recyclability, and as such develop a Carpet Stewardship Program. As part of this program, C.A.R.E. conducts a national survey of carpet recycling in the United State and Canada, which it believes represents more than 90% of the volume of carpet diverted and recycled in these two countries (Figure 5.2). Based on its surveys, it appears that more than 97% of the postconsumer carpet material processed was used in the United States with 58% recycled into engineered resins; 27% into new carpet, as either carpet fiber (20%) or carpet backing (7%) [11].

Operation Green Fence

Asia, principally China, has been the major export market of plastic recyclables from North America and Europe. In recent years, the abundance of nonrecyclable materials within bales shipped to China has become a major concern of Chinese officials. In 2013, China announced an inspection program, titled "Operation Green Fence," which requires a special team of custom officials to inspect nearly every container reaching Chinese ports [10]. There is a zero tolerance for banned items (e-waste, green waste, human waste, food waste, medical waste, etc.) and a preinspection policy in US ports before these commodities are loaded onto ships before shipment. Because these operations slow down port operations, shippers are now seeing rising costs to hold containers until they are inspected. It is currently unknown that what the market impacts will be on prices paid for these commodities and the further development of domestic recycling markets in the United States and Europe.

Scrap Tires

Based on industry statistics, nearly 60% of all the rubber consumed in the United States was used to manufacture passenger and truck tires [12]. In 2011, there were 287 million tires generated (5.2 million tons), approximately 85% for replacements of used tires and the remaining 15% shipped with new vehicles [12]. Averaging almost one scrap tire per person per year in the United States, the vast numbers of these scrap tires pose enormous difficulties for local government recycling programs.

Markets for scrap tires include use as tire-derived fuel, which is a low sulfur, high heating fuel, artificial reefs, playground equipment, mulch, carpet padding, tracks for athletic fields, and uses in road embankment or roadfill.

E-Waste

As more and more businesses and consumers move into the electronic age, there is more electronic waste, e-waste, that is produced with equipment upgrades. These include desktop computers, laptop computers, computer monitors, printers, and televisions. According to a 2011 survey undertaken by the Institute of Scrap Recycling, the US recycling industry processed 3–5 million tons of used e-waste equipment [13]. Estimates are that 70% of this equipment is turned back into scrap steel, aluminum, copper, lead, glass, plastics, and precious metals such as gold and silver.

In recent years in the United States, there has been a steady increase in the number of states with laws on electronics recycling. Currently, there are 25 states with program laws covering about two-thirds of the US population. In all these states, except for California, which is funded by an advance disposal fee, manufacturers are responsible for establishing and funding these e-waste programs. California's program requires sellers of covered electronics to add a fee of $6–10 to the price of the product and remit that fee to the state. Recyclers then bill the state for the costs of collection and recycling for devices that are returned by state residents.

A continuing controversy in the e-scrap industry has been exports of used electronics to the Developing World. Governments, NGOs, industry, and the general public have become sensitized to media stories of smoldering "e-wastelands" in China, India, and West Africa where low-paid workers are bent over discarded e-waste and toxic materials are disposed of in nearby bodies of water [13]. With little information available, debate has been largely focused on the environmental damage from exports of e-waste from the affluent northern countries to marginalized peoples of the global south. Regulations have been enacted in many countries banning the export of e-waste in favor of local processing facilities. Unfortunately, these outright export bans negate the fact that many e-waste processing steps still require manual processing, and labor in the Developing World is still infinitely less expensive. A study released by the United Nations University in 2013 concluded that as long as the profit potential of discarded e-waste is constrained by strict environmental standards in the north, excessive taxes, and high labor costs, it will be difficult to stop the flow of e-waste to the Developing World [13].

Yard Waste

Yard waste is one of the largest contributing components of the MSW stream. By weight and volume, it is only exceeded by paper products. Nationally, it represents 15–20% of the solid waste stream. By definition, this waste is generated from the homeowner's yard, tree and brush trimmings, garden waste, grass clippings, and leaves. The majority of this material is generated on a seasonal or cyclical basis. Tree trimmings enter the waste stream normally in the late winter or early spring, while garden waste and grass clippings are produced during the spring and the summer. Yard waste is also generated by landscape maintenance firms from the upkeep of both residential and commercial properties.

For yard waste to be recycled from the homeowner, source separation is an absolute requirement. Material is best prepared at the source. Tree or brush clippings should be bundled, garden waste should be put in containers, and leaves put at curbside in bags or loose for vacuum pickup. Bagging leaves, garden waste, or grass clippings creates problems for the composting process with an additional processing step required to remove the bags. Plastic bags are a contaminant and interfere with the decomposition process of composting. Furthermore, residual pieces of bags in the finished compost product are esthetically undesirable and lowers the market value.

Some of the materials would have to be processed prior to composing. Brush, woody materials, and garden waste need to be grinded to provide proper size and surface area for microorganisms to work on. Leaves generally do not need to be grinded, although grinding does accelerate the digestion process.

The market for organic compost is not as developed as that of the commodity recyclable materials. Typical markets for compost are greenhouses, nurseries, golf courses, landscape contractors, turfgrass farms, cemeteries, agriculture, and topsoil suppliers. It is also used as landfill cover, along roadsides for erosion control, and for other nonfood horticultural activities. Most often, yard waste composting projects realize minimal to zero revenues, claiming benefit from the avoided cost of landfilling if the product is taken away by an end user or from the avoided cost of topsoil if the product is used as landfill cover.

Markets for compost can be divided into four major categories which include growers, processors, wholesalers, and bulk users. Growers can include golf courses, nurseries, landscape companies, lawn maintenance businesses, and sod and sod service businesses. Processors refine the compost to meet specific conditions and market the material. Processors may include topsoil companies, fertilizer companies, or sand and gravel companies. Wholesalers typically purchase and sell bagged compost at garden supply or nursery type businesses. Bulk users utilize large quantities of compost to cover acreage for land reclamation, landfill cover, or for park or roadside stabilization after construction. Since most road construction is done by state Department of Transportations (DOTs), discussions with DOT officials could result in a significant market.

References

[1] American Forest and Paper Association, <www.afandpa.org>. Accessed April 16, 2013.

[2] Newspaper Association of America, <www.naa.org>. Accessed May 2, 2013.

[3] Miller C. Profiles in garbage, Waste 360, <http://waste360.com/print/old-newsprint-onp/profiles-garbage-newspapers>. Accessed May 16, 2013.

[4] American Forest and Paper Association. 2012 AF&PA sustainability report; 2012.

[5] Glass Packaging Institute, <www.gpi.org>. Accessed May 17, 2013.

[6] Moore Recycling Associates. Plastic recycling collection national reach study: 2012 update, prepared for the American Chemistry Council; 2013. Accessed May 17, 2013.

[7] Verespej M. Polypropylene: the next big thing in recycling, PRU; 2013. p. 12–20. Accessed May 21, 2013.

[8] Linnell J. Signal to noise, Waste Age 2010. p. 60–64.

[9] Moore Recycling Associates. 2011 National post consumer plastic bag and film recycling report, prepared for the American Chemistry Council; 2013. Accessed April 4, 2013.

[10] Plastics Recycling Update. April 2013 volume 26, No. 4.

[11] CARE. 2011 Annual report, carpet America recovery effort, <www.carpetrecovery.org>; 2013. Accessed April 9, 2013.

[12] Miller C. Profiles in garbage, waste 360, //waste360.com/waste-facts#tires0.com/print/tires. Accessed April 8, 2013.

[13] Salehabadi D. Solving the e-waste problem, United Nations University/StEP Initiative; 2013.

[14] US Environmental Protection Agency. Municipal solid waste generation, recycling and disposal in the United States: facts and figures for 2010, <http://www.epa.gov/wastes/nonhaz/municipal/pubs/msw_2010_rev_factsheet.pdf>; 2011. Accessed April 8, 2013.

6 Public Education Programs

How Do People Learn

Why should it matter? Given the rarity of public education funds, efficiency and rate of return on each dollar invested is critical. The most effective programs are responsive to human psychological needs for knowledge development. Before one begins to plan for recycling education, it is important to understand how people accumulate knowledge or form patterns of behavior. While this book was not intended to serve as a psychology primer, it can suggest key points, which if understood, could influence the direction taken with public education efforts.

Two Basic Approaches—Piaget

Two basic approaches worth considering begin with a decision as to when and how to begin public education. If you subscribe to the belief that knowledge is best built on a "blank slate" or a fresh foundation free of fissures or foreign elements, then you might adopt an approach that reflects the philosophy of the renowned developmental psychologist Jean Piaget and his theories of cognitive development. Put simply, Piaget believed that knowledge is the product of a series of sensory building blocks called schema [1]. The schema contains bits of information that we gather through physical and emotional experiences.

To relate schema to recycling, most people in their late teens or older when handed an empty tin can and asked to identify it would respond that it is a piece of garbage. Logic would follow that in most cases, in the home; they would treat an empty can as though it were garbage and throw it in the trash. On the other hand, people who have been exposed to recycling from a very early age when asked the same question might respond that the can is something that can be recycled. In the home, they are likely to respond by tossing the empty container in the recycling bin or bag.

For communities interested in and willing to begin education with the very young, a Piaget-oriented approach would work well. Teaching children, from the beginning of their educations (as early as age 2) about recycling and of ways that they can participate, can set the stage for lifelong behavior. Obviously, beginning with very young children is a departure from many public education approaches, yet has the potential for long-term reward. That reward would be realized as the community's youngest residents grow older, influence their families and friends and form the basis of a permanent behavior pattern.

Solid Waste Recycling and Processing. DOI: http://dx.doi.org/10.1016/B978-1-4557-3192-3.00006-3

Two Basic Approaches—Skinner

The second approach involves public education designed to change habits already in place. Some would look toward lessons taught by psychologist B.F. Skinner and his work in applied behavior analysis [2]. Skinner's "science of behavior" maintains that all behavior is the result of a natural sequence of events. The events consist of an ANTECEDENT which yields a BEHAVIOR and then a CONSEQUENCE.

$$\begin{matrix}\text{Antecedent} \\ (\text{prompt or cue})\end{matrix} \rightarrow \begin{matrix}\text{Behavior} \\ (\text{reaction})\end{matrix} \rightarrow \begin{matrix}\text{Consequence} \\ (\text{reward or punishment})\end{matrix}$$

Related to recycling, the antecedent could be an article in a newspaper about mixed-paper recycling. The article would prompt the realization that the reader has a choice of behavior when it comes to the disposal of mixed paper. A recycling behavior might be the accumulation of junk mail or other specified mixed-paper and periodic delivery to a drop-off facility. The consequence could be a financial reward provided by the drop-off center or simply an "atta boy" from the worker that runs the facility that makes the individual feel good about the visit. The "atta boy" is positive reinforcement which will be recalled the next time mixed paper becomes an issue in the home. The resident is likely to repeat the recycling behavior if the reinforcement was positive enough. Conversely, if upon visiting the recycling center the resident gets a flat tire due to debris on the site, the negative consequence that accompanied the visit will probably discourage this particular recycling behavior in the future.

If you plan to change the behavior of the people serviced by your program, it is important to design the program around clearly defined recycling initiatives, intended residential behavior, and positive consequences. A good example would be a program that seeks to recover type 1 plastic containers. Your public education effort must reach out to those that generate this type of material in their homes or businesses. Clear direction must be provided to enable residents to determine the difference between type 1 and type 2 containers, how you need them prepared and method through which they are stored and transported to the curb for collection. By clearly describing the activity and desired resident response, you enhance the odds of the behavior leading to the successful collection of the material. If there is a misunderstanding and plastic bottles are rejected at the curb, it will take but a few of these experiences to take the form of punishment rather than reward. If the resident feels that recycling efforts will result in punishment, the result could be a drop in participation.

Communication Do's and Don'ts

Research on motivating desirable waste management behavior conducted by Geller and Lehman and published in the *Journal of Resource Management and Technology* identified the following do's and don'ts [3]. Regardless of education philosophy, these guidelines apply in most areas of motivational communication. By taking note of them, you will likely avoid heated debate with residents or elected officials in the future:

- Motivational signs, messages, or slogans are usually not sufficient to precipitate changes in behavior, unless the desired response is viewed as convenient and the cue occurs at the point of the desired action.
- General requests or recommendations may be inexpensive to field but could be useless in changing specific behavior. Example, "let's all recycle" versus "Please participate in our community recycling program by placing your newspapers, glass containers (clear, green, and brown), number one and two plastic bottles and aluminum cans in the city provided, recycling container, at the curb the first and third Wednesdays of the month."
- Educational incentives are usually more effective when linked with audience participation via verbal exchange or hands on demonstration. A good example of an audience participation approach could be a material sorting game or a demonstration where children assist an instructor in making paper out of waste material. Lectures or video presentations that rely on passive listening are less effective.
- Strategies that set examples or provide role models for changes in behavior provide significant return on investment, especially when linked with support material echoing the principles of the example. One of the most common approaches could be called "practicing what you preach." Many communities adopt recycling within the administrative and operations branches of government. They adopt and promote polices which encourage recycling and waste prevention by government employees and establish guidelines for purchasing recycled or recyclable materials.
- Public commitments and goal setting approaches are very effective and, as communications strategies, inexpensive to use. Many school programs include the use of recycling pledge cards. These are used to solicit written commitments from students. One of the most common examples is the Recycling Ranger program used by many schools in New Jersey. After attending a recycling education program, children are made deputy Recycling Rangers (certificates or "membership" cards are provided).
- Strategies that gives considerable attention on the consequences of behavior, either reward or punishment, must be accompanied by clear directions as to the recycling initiative that prompts the behavior. If a promise of reward is made, yet not provided due to a misunderstanding of the recycling rules, program credibility may be lost or retarded until proper instruction is provided.
- Prompts with demanding instructions or unpleasant consequences should be avoided. The threat of negative reinforcement often yields behavior opposite of that desired. Recyclables rejected at the curb often result in hard feelings and resentment that are difficult to reverse.
- Individual perceptions of the type of reinforcement (positive or negative) may determine the impact on behavior and require careful consideration before committing to full implementation. If an elaborate positive reward is offered such as payment for material recycled, a long-term commitment may be required, regardless of market conditions in order to yield a long-term change in waste disposal behavior. The sponsoring entity needs to carefully consider the true cost of such a program.
- Reinforcement linked to group behavior yields an additional level of social control and is desirable where feasible. Peer pressure is very effective when it comes to recycling. A good example is the impact of being the only house of the block to not put recyclables out for collection. There is usually a noticeable absence of participation the entire neighborhood can see.

- Perceived threats to freedom and negative psychological responses can be avoided by participatory planning approaches. Many programs are planned with the input of a public advisory group. Often referred to a Solid Waste Advisory Committees (SWAC), the SWAC is made up of individuals who reflect the businesses and residents of the community. The SWAC, media, and periodic public hearings will help to avoid misconceptions or feelings of disattachment with the recycling process.
- Reinforcement should be provided on a long-term basis and seriously enforced until residents adopt recycling behavior as a natural part of their routines. The goal is to assist the public in forming a recycling habit. That can be more difficult than it sounds because, generally, recycling habits will require a change in solid waste management habits that have been in place for the lifetimes of local residents.
- A specific entity should be clearly assigned the responsibility and resources needed to provide reinforcement for recycling behavior over an extended period of time. Often referred to as the local recycling coordinator of solid waste director, the individual(s) assigned must be given clear goals and objectives as well as the resources needed to reach levels of expectation.

Experienced recycling coordinators could contribute additional do's and don'ts to this list. Probably the most important thing that can be done is to design public education programs specifically to meet the needs and perception levels of the host community. Conversely, many communities have made costly commitments to recycling education and operational approaches on a "copycat" basis. The belief that money can be saved by adopting the neighboring community's program is an invitation for disaster. Your neighbor may have copied its program from somewhere else with conditions entirely different from yours.

Where Do You Begin?

Before making plans or commitments to recycling education strategies, it is important to determine the level of public awareness that exists within the community. Many recycling coordinators have assumed that residents either do not know anything about recycling or know and care as much as they do. Rarely will either scenario be true.

The Monmouth County Example

A good example can be found in Monmouth County, New Jersey. Monmouth has a population of nearly 600,000 and is located on the New Jersey shore about 30 miles from New York City. In the late 1980s, work began on a mandatory residential and commercial recycling program that included all 57 communities within the county. Careful waste stream studies were conducted over a 6-year period beginning prior to the mandatory program and continuing over a period during which most municipal programs matured.

In the early 1990s, decisions were made to reassess the progress of the recycling program. One of the tools used to conduct the assessment was a professionally designed and administered public opinion survey [4]. The survey revealed a surprising lack of awareness of the problems facing the county relative to solid waste management and its residents' responsibilities relative to being a part of a solution. This response was ironic in light of an extensive, long-term educational effort at municipal, county, state, and national levels. Further, for nearly 2 years, the county was the site of heated debate relative to the construction of a material and energy recovery facility. Plans for the waste-to-energy plant were ultimately canceled due to a loss of a referendum.

Upon review of the survey data, county officials changed course with their plans for a new wave of public education. Specific elements of the county population were targeted for additional attention, and the overall structure of the recycling and waste management message was modified to fill critical gaps in knowledge and awareness. There was initial concern relative to the cost of the survey (approximately $20,000) and the time required to plan and conduct the program (4 months). Of greater concern was the potential greater loss of funds and time had the program not been effectively targeted and the best message developed.

Many recycling coordinators will be faced with evaluating the results of past or ongoing public education efforts. While it may appear to be cheaper to guess or speculate as to effectiveness, in the long run it pays to begin with a properly designed and administered survey.

Public Opinion Survey

The Monmouth County Public Opinion Survey is a good example of an effort specifically designed to determine awareness, attitudes, and opinions toward municipal solid waste issues including waste generation, reduction, recycling, and disposal. The results of the study were used as the basis for the development of a communications plan for implementation by the County.

While designing a survey for any community, it is important to remember that recycling is a major element of a larger effort to manage municipal waste. Given the time and expense needed to implement a survey program, the tool should be used to assess perceptions and attitudes on a broader basis. For example, it would be tempting to gear the survey strictly to the measurement of public participation rates. By doing so, an opportunity would be lost to determine attitudes relative to waste reduction.

Objectives

As you plan your survey, it is helpful to first establish a clear set of objectives. Before they are finalized, share them with other individuals involved in recycling efforts and solid waste management in your jurisdiction. Make sure that key elected officials are aware of your plans and have an opportunity to contribute suggestions relative to the objectives.

Your survey should include the determination of the following:

- The awareness level of existing solid waste management disposal problems and possible solutions.

- The level of knowledge of current solid waste management initiatives underway in the community.
- The current level of mandatory and voluntary resident recycling and/or composting efforts.
- Awareness and attitudes toward waste reduction options.
- The level of perceived public responsibility for waste generation and management activity.
- Attitudes toward recycling options including levels of participation and willingness to take on additional responsibility.
- Where significant differences in opinions exist, to allow for the development of a communication plan designed specifically to improve overall public participation and performance in waste management programs.

Methodology

Once the objectives have been identified, turn to the development of methodology for the survey. Care must be taken at this point not to mislead yourself as to the approach planned and the value of the subsequent results. Carefully consider retaining an experienced market research firm or other entity skilled in scientific polling and statistical analysis. Sample size will be critical to interpreting the reliability of the survey results. The market research firm can guide you through the development of the sampling approach and help you tie the results to a level of confidence and the margin of error based upon your sample size. Often random sampling is employed based upon telephone numbers from local directories. Generally, a basic inventory of numbers will be drawn, a portion of the first digits will be retained, and random endings added. This approach takes into account unlisted and new numbers. Callbacks are administered during different times of day and evening to eliminate time-of-day bias. It is also helpful to build in quotas to the sample selection process. Type of housing unit, age, education, income, and gender should also be considered. Care should be given to assure realistic splits for housing ownership versus renting, single-family homes versus multifamily units surveyed, age across a number of age ranges, multiple categories for education, income on a broad range and gender.

Analysis

As the analysis is conducted, it is important to determine frequencies of responses for all questions. Cross tabulations should be done according to demographics to determine significant differences in attitudes and to identify significant patterns.

After the data is analyzed, look for major differences in responses to the questions. For example, the Monmouth County survey analysis focused on differences relative to the following key issues:

- Opinions on garbage
  - Who creates the most garbage in Monmouth County?
  - Who has the most responsibility for solving the garbage problem?
  - Should education on recycling and garbage disposal be included in the school curriculum?

- Awareness of waste disposal problems
  - Do you think Monmouth County has a problem in disposing of its garbage?
  - Would you suggest ways of reducing the amount of garbage thrown out in Monmouth County?
  - What are some other good ways of reducing garbage?
  - Are you aware of Monmouth County-sponsored mixed-paper recycling centers?
  - Where do you get the most of your information about recycling?
  - What materials are residents required to recycle in Monmouth County?
- Behavior with regard to recycling
  - What materials (mandatory) do you currently recycle?
  - What materials do you voluntarily recycle?
  - Are you using the mixed-paper recycling centers?
  - How are you willing to participate?
  - Additional materials on a voluntary basis?

Results

The conclusions of the survey will reveal the level of awareness within the population surveyed. It will provide evidence as to the need for public education and in what areas (demographic and programmatic). It will determine awareness relative to specific program elements, and participation rates can be quantified. The survey can provide information that will serve as the foundation for waste reduction efforts. You will be able to provide a quantitative assessment of how well the existing programs/facilities are doing and identify where more work is needed.

Armed with survey results, you can target your communications plan to the actual needs that exist within the community. At the same time, you can use the findings to help in selecting the most cost-effective media through which to deliver your message, the appropriate format, and frequency. Your communication program will adopt a "rifled" approach versus a less-effective "shotgun" marketing program.

The Communications Plan

The public education or communications plan is the road map to implementing your strategy. It should identify a target audience, the specific actions desired, and desired responses. It will include a mission statement, a central theme, measurable goals and objectives, specific strategies including action items, schedules, and budgets.

Media buys should be clearly identified including specific media form (i.e., print, radio, cable television), frequency, and budgets. As part of the plan, provisions should be made to conduct periodic follow-up surveys to gauge results and guide adjustments to the plan.

The budget should be broken into specific educational elements and the total budget conveyed on a cost per resident basis. Many programs are dependent upon free coverage in newspapers or through public service announcements on radio or television. Given the need for sustained educational information, dependence on free coverage may yield less than desired response. Funds should be set aside for paid advertisement if your program is focused on print or radio media strategies.

Large sums of money are not necessarily a requisite for effective public education. Some plans can be geared toward the custom adaptation of materials and strategies developed by others. Also, as mentioned under do's and don'ts, strategies based upon modeling or hands-on experiences are cost-effective and can result in favorable results. As you develop a plan, pay close attention to the information your survey provides and avoid reinventing the public education wheel. There are lots of high-quality materials available for a fraction of their original development costs.

Public Education Strategies

Your communications plan will be like a pie divided into slices or substrategies, each aimed at achieving the mission you identify in the beginning of your plan. Hundreds of substrategies have been developed by talented educators and recycling coordinators from around the country. Attend any conference on recycling or solid waste management and you will see new examples of classroom initiatives, the innovative use of videotape, mailings, poster contests, county fair displays, public service announcements, paid advertisements, recycling logos, characters, and clever costumes. You will also learn about new or innovative waste reduction strategies, information sharing networks, trade journals, and free materials available via manufacturing organizations that represent paper, plastic, steel, ascetic, and glass packaging interests. Free advice is also available through the National Recycling Coalition, the Solid Waste Association of North America, and others.

In this section, three detailed examples of public information programs are offered. Each represents a comprehensive strategy and examples of substrategy. Each is either unusual or an example of a particularly effective approach.

The Clean Community System

Keep America Beautiful, Inc. (KAB), a nonprofit organization originally founded to address litter prevention and community involvement in solid waste issues, developed the Clean Community System (CCS) in the mid-1970s. The system has proven to be one of the most effective approaches to changing human behavior on a community-wide basis. KAB has based its model on the promotion of practical collaboration between local government, business, and civic organizations aimed at solving solid waste management problems. The CCS has been implemented in more than 400 US communities in nearly every state [5]. In addition, it has influenced litter-control programming in six other countries: Australia, Bermuda, Canada, Great Britain, New Zealand, and South Africa. The CCS model provides one of the best working models for organizing, motivating, and evaluating community action toward the large-scale solution of environmental problems.

Motivational strategies, particularly the use of heavily promoted awards programs, are important to the CCS approach. Positive reinforcement via social approval and public recognition are the primary rewarding consequences. The achievement of "Certified Clean Community" status is a major potential motivator for mayors, municipal staff, and volunteers.

The Resource Community System (RCC) proposal is a two-stage plan through which communities initiate waste management procedures including recycling and waste reduction in order to earn eligibility as a "Certified Resource Conserving Community" (CRCC), and then specific criteria are developed for qualification as a "Responsible Resource Community" (RRC).

As under the KAB plans, completion of each stage would result in heavily publicized praise and if possible, financial rewards through government and industry grants. Positive reinforcement must be applied at every stage of the process in order to encourage progress to ultimate recognition and the achievement of the mission. Additional incentives will be needed in order to maintain performance levels after the community has achieved RRC status. From a political perspective, elected officials may strive to be certifiably responsible with local resources. The potential for this model extends far beyond recycling to responsible government and conservation.

The education process begins with a letter of endorsement from a mayor, county executive, or other ranking elected official. With that letter comes a commitment to send a work team to an RCC training conference and to provide sufficient funds for the first year of operation. Training conferences can be planned and executed via a county, state, or university extension system. Under the KAB model, local training entities are accountable to a national coordinating organization. The work team representing the community is made up of individuals from government, commercial and community organizations, and related volunteer groups.

At the training sessions, the community work teams learn about methods for coordinating, motivating, and evaluating a local program for recycling and/or solid waste management. The team assembles its own plan designed to address the needs of the residential/consumer, governmental/institutional, and commercial/industrial sectors. During workshop training, attendees learn about waste prevention, reduction, recycling, and proper waste disposal. The KAB workshops focus a significant amount of attention on litter control. During the same workshops, attendees are asked to work with RCC team members to develop an incentive/reward program for prompting and encouraging long-term efforts in the desired behavior.

After a week of training in the workshops, the work teams are asked to implement a large-scale program in their home regions. Communities that participate in the workshop training are certified as "Resource Conserving Communities." Progress by the local work team is monitored and rewarded by local, state, and federal entities. The ultimate goal is that of recognition as a RRC. Prior to achieving RRC status, the RCC work team must establish a local headquarters; receive funding commitments; present RCC workshops to key local leaders in government, institutions, commercial activity, industry, and community agencies. The RCC must select demonstration projects representing local needs and organize an RCC Action Council.

Under the KAB model, the Action Council is essential from both an organizational point of view and as a source of workers for implementation of the RCC plan. Typically, members of the council would come from the residential/consumer, government/institutional, and commercial/industrial sectors. In addition, individuals with backgrounds of litter control, recycling, economics, education, engineering, psychology, and sociology would also be recruited. Priorities for the council would include the establishment of a resource center for information relative to recycling and solid waste management; provision of a communications link to local experts who administer the RCC training programs, offering certification as RCCs and RRCs; organizing and monitoring the RCC action teams responsible for waste management education, promotion, and demonstration projects; evaluating progress of the RCC action teams; and developing criteria for an RRC.

KAB suggests that the Action Teams be organized as follows:

- The Communications Team would be responsible for publicizing the activities of the RCC, providing for positive reinforcement of individual and group activities' efforts to encourage resource conservation and other efforts that increase recycling. All communications media should be used and an internal newsletter developed.
- The Education Team would be responsible for keeping all other teams informed of emerging issues or techniques for solid waste management and for promoting recycling in public schools. It may also sponsor public education programs for teaching waste management principles including waste reduction, recycling, and reuse techniques.
- The Community Team would be responsible for locating and coordinating local civic organizations in resource conservation and recycling initiatives. This committee would train civic volunteers to teach workshops and initiate neighborhood projects and assist in evaluating community programs. It would also conduct award programs to provide additional positive reinforcement for individuals and groups working toward RCC goals.
- The Government Team is responsible for examining government policies and encouraging those that promote sound planning and resource conservation efforts over a long-term basis. This team could suggest policy, programs, and legal initiatives aimed at motivating waste reduction, recycling, and responsible disposal practices. The Government Team would also plan and execute government workshops on waste management issues for public employees.
- The Industrial Team is charged with promoting waste reduction, recycling, and sound disposal practices in the commercial/industrial sector of the community. It should establish prominent examples of practical and appropriate waste management practices via the private sector. It should develop and sponsor workshops on waste reduction, recycling, and waste management for groups in the commercial/industrial sector.

While the KAB model provides a comprehensive framework on which hundreds of successful programs have been structured, the most important factor to the success of this model has been the coordinated approach through which it is administered. Each waste generation sector is recognized, provisions are made for training, follow-up measurement and significant emphasis placed upon positive reinforcement. Roles are incorporated for multiple levels of government and the use of existing mechanisms like cooperative extension services and state universities to assist in educational efforts. KAB was successful in implementing its model on a national scale. Many recycling coordinators have access to local KAB affiliates and their trained professional staff, as well as extensive education resources available at little cost.

Beginning with Preschoolers—The Mister Rogers Model

There are at least two major factors that can influence a community's ability to reach higher recycling goals. The first is the community's ability to align programs with recycling markets such that those markets will be able to absorb recovered material over an extended period of time. The second factor is public attitude and willingness to participate in the solid waste management program on a long-term basis. In many cases, public attitude must be compatible with dramatic increases projected in recycling and reduction efforts over the long-term, 20-year planning period.

This example, which reflects Jean Piaget's theory discussed earlier, was designed to take advantage of a "clean slate" approach to education. By focusing on very young children, aged 2–4, a foundation is constructed for behavior responses that can last a lifetime.

Preschoolers between the ages of 2 and 4 are ready to begin learning about recycling and waste management. In 1989, public television's MISTER ROGERS' NEIGHBORHOOD featured a five-part series of programs on the environment [6]. One of the programs (No. 1617) was specifically tailored to convey a message on recycling and conservation. The 1989 effort was the first major initiative aimed at carrying a recycling "lesson" to very young children. Aired on nearly 300 stations of the Public Broadcast Service, more than 14 million households tuned in to view the program. Since its debut on Earthday of 1989, the series of programs has been broadcast nationwide 10 times.

Briefly, the program opens with Rogers walking into his house carrying a bag filled with glass bottles and cans. He talks about recycling and sings, "I like to take care of you." His friend stops by and helps him separate recyclables into two boxes. Together, they visit a recycling center where they talk with the workers and watch what happens to the bottles, cans, and other material. Upon returning home, the program transits to a puppet segment where the discussion shifts to a local garbage problem. The characters begin to search for a new landfill and some suggest that recycling would help the situation.

In the final program segment, Mister Rogers asks his viewers to think about what we really need before we buy and to think about other things they can do with waste material before we throw them away. This final request has major significance. Preschool children love and respect Fred Rogers. A poll by Playskool asked preschoolers in five American cities which famous person should be the next president, 45% said Mister Rogers. If you share Fred Rogers' opinion that "attitudes are caught, not taught," you might conclude that the program described above provides a broad reaching opportunity for millions of Americans to "catch" a recycling and waste reduction attitude.

Research by several researchers would support a theory that the widespread application of the recycling and conservation message contained in the MISTER ROGERS segment could have a major impact on local behavior. The researchers used 10–15 min segments from MISTER ROGERS' NEIGHBORHOOD to test groups of small children [7]. They found that after viewing the clips, the children were found to have improved in task persistence, self-control, and tolerance of delay in comparison to children who viewed either antisocial or neutral films.

The use of the program as part of a nationwide, comprehensive public education strategy could yield significant, long-lasting results. Our research concluded that if a national effort were launched in 1990, by the year 2010, over 76 million Americans or 27% of the nation's population could have been exposed to a positive message relative to recycling and conservation. Even under the most ideal circumstances, the program could not reach every child directly. Language, location, economic resources, and parental attitudes pose specific obstacles. The residual effect of the program could help to offset the obstacles.

Public attitude is almost certain to be influenced by such an initiative. Within this century, society has shifted its approach and attitude relative to waste disposal. Not long ago, a parent would ask a child to take out the garbage. The son or daughter complied by carrying the waste to the front room window and dumping it onto the ground below. Within the next few years, assuming significant strides are made in public education, when a father makes the same request, we suspect that the child will automatically spend more time and effort preparing the waste to enter the management cycle.

How do you take this example and apply it on a local basis? If used to the greatest advantage, the MISTER ROGERS' material including the videotape and an accompanying 24-page activity guide could provide the comprehensive strategy needed for long-term results. The substrategies could include a publicity program entitled "Starting at the Beginning," focusing attention on the need to begin good waste management and conservation habits early on. Publicity could promote frequent local broadcasts on public access television (with specific permission from Family Communications Inc.). It could promote the use of the video and workbooks in local preschools, kindergarten, and first grades as well as library story hours. The guides could be distributed to homes with young children. Poster competitions can be designed around the "Starting at the Beginning" theme as could advertising campaigns.

Each of the above-mentioned substrategies has been used by communities across the country. Aside from the obvious educational benefits, sponsoring communities benefited from the high level of name recognition that accompanies the material and by the fact that the material is available for any noncommercial use to the general public.

Varieties of Educational Methods

One of the most popular outlets for recycling education is the public school. In 1991, more than 148,000 California students from the state's 15 most heavily populated counties were the focus of a 2-week recycling education program. According to Roger O. Scott, head of the Southwest Regional Laboratory (SWRL), who developed the recycling education program, almost 5000 teachers and 232 public and private schools used the curriculum materials [8]. About 60% of the participating teachers taught kindergarten through third grade. The program included a series of ten 25 min lessons emphasizing four objectives: the reasons why recycling is important, the kinds of containers that can be recycled and redeemed, collection procedures for containers, and locations for recycling. Students were given color activity booklets and they listened to a tape and viewed a film strip explaining recycling. Beyond the structured program, teachers and students in 95% of the participating schools collected recyclables—as an opportunity to apply what they learned.

In an attempt to measure the results of the initiative, SWRL conducted a survey with about 75% response rate. SWRL reported that more than 94% of teachers indicated that their students enjoyed learning about recycling and more than 92% said that their students became more aware of recycling. Parents surveyed agreed that their children enjoyed learning about recycling. According to Scott, 80% of the parents surveyed reported that their child recycled more and 70% said that their families are now recycling. There is little doubt that children carry the recycling message home.

While the California results were impressive, recycling coordinators should keep in mind that getting teachers interested in recycling education can be a problem. Teachers are busy and often overwhelmed by the number of requests and mandates they receive to include specific instruction in an already crowded school day. Some teachers may not have the time within or outside of the classroom to research new topics and gather activity material. Tight school budgets present an additional constraint that may even preclude the cost of material reproduction.

As you plan to gain access to local students through the local school system, make sure that you provide as complete a package as possible to teachers. It should be age appropriate, of a professional quality, and easy to read. For items intended for distribution, a sufficient number of copies should be available. They should be sensitive to a need for ethnic balance and other socioeconomic concerns. Finally, do not forget that your material will have to fit into a specific time slot, probably no longer than one and half hour.

The examples mentioned earlier in this section have provided a broad range of applied approaches. It will be helpful to consider the following general list of substrategies as you prepare community specific programs.

A survey conducted by the Environmental Hazards Management Institute (EHMI) in Durham, New Hampshire, identified 30 educational methods, broken down into four general categories [9]. While some may be familiar, there are sure to be a few that prompt new strategies for recycling coordinators (Table 6.1).

One important and potentially fruitful area not revealed by the EHMI survey has to do with the use of public access television and the availability of production facilities and air time via most local cable television stations.

Social Media

Within the last few years, many solid waste agencies worldwide are utilizing social media as a means to interact with new audiences and sharing valuable information and data. "Social media" is defined to be the web-based and mobile technologies where things such as messages, videos, blogs, and pictures can be shared with various communities (Table 6.2). These social media technologies are oftentimes used to increase the agency's solid waste and recycling programs to a much wider audience than would otherwise be reached with traditional print and electronic media.

Table 6.1 Methods for Recycling Education

Print media	Newspaper announcements
	Newspaper advertisements
	Newspaper articles
	Bumper stickers
	Recycling center signs
	Recycling truck signs
	Other signs
	Newsletters
	Mass-mailing of brochures
	Publicly available brochures
	Mass-mailing flyers
	Publicly posted flyers
	Recycling bin stickers
Electronic media	Television advertisements
	Television interviews
	Radio advertisements
	Radio interviews
	Telephone hotlines
	Social media (Facebook and Twitter sites)
Other	Recycling surveys
	Recycling committees
	Recycling awards
	Waste reduction curriculum
	Other (as specified by community)

Source: From Ref. [9]

Table 6.2 Social Media Categories

Communication	Blogger
	Facebook
	Linkedin
	MySpace
	Twitter
Collaboration	Delicious
	Digg
	Reddit
	StumbleUpon
Multimedia	Flickr
	Upstream
	Vimeo
	YouTube
Other	IPad/IPhone Apps
	Podcasts
	RSS Feeds
	Widgets

Source: From Ref. [10]

Increasingly, it has become a method of choice in many locations to interact with younger generations who spend more time on the Internet. Also, it has been used to target information to defined communities.

Web Sites

Web sites have become a very important and key way to publicize waste reduction and recycling, and are presented in a variety of forms and methods very much dependent on each individual city history, procedures, and focus. A key element is the commitment and energy put into the program by elected officials and department heads in assisting with coordination between multiple departments with overlapping solid waste responsibilities and interests. The web site designer needs to be well informed and provided with continually updated information on needs and priorities, as well as the ability to design and illustrate the programs.

Information on many other city programs is provided in other report sections and indicates varying amounts of information available. Several of these cities could be models for Skagway to use in updating its web sites.

Print Material

Other forms of print material can also be effective forms of education and outreach and can include the following:

- Billing stuffers
- Direct mailings
- Phone book section insert.

These avenues of advertising are most effective for permanent residents of the community.

Signage

Taking advantage of colorful signs displayed around town can be effective in communities that have a large transient population or to those who host large numbers of tourists. Establishing a recognizable mascot or theme that is found at different venues frequented by this population can help encourage recycling participation. Establishing clear messages as to what is and is not recyclable also helps in fostering an increase in source separated recycling in public areas (Figure 6.1).

Events and Awards

Making recycling a part of the planning and implementation of all public events is a good way to increase awareness beyond routine information. Similarly, developing award programs, such as a green business recognition program, are other ways to build recognition and "branding" for recycling.

Figure 6.1 Signage used at Seattle airport.

Technical Assistance to School and Businesses

Volunteers or paid recycling coordinators are instrumental in providing technical assistance to teachers and administrations at schools and to managers and employees at local businesses. Engaging these sectors and providing useful instruction and incentives to them can greatly enhance a community's recycling program. Additionally, these sectors provide a viable population for embarking upon pilot studies for various recycling programs, such as a food waste diversion program.

A good example of this type of program has been implemented by the City of Irvine, California [11]. In July 2007, the City of Irvine passed a resolution adopting Zero Waste as a long-term goal for the City, in order to eliminate waste and pollution in the manufacture, use, storage, and recycling of materials. Reaching this goal will involve the active encouragement of residents, businesses, and agencies to use, reuse, and recycle materials judiciously, in addition to encouraging manufacturers to produce and market less-toxic and more durable, repairable, reusable, recycled, and recyclable products.

The first project undertaken under this directive has been an outreach and technical assistance program for food service establishments in the City. This project worked with restaurants to promote Zero Waste concepts and identify ways to help them reduce waste and work toward Zero Waste.

The City developed a community-based social marketing approach to promoting Zero Waste through the utilization of motivators and reduction of barriers identified during a preproject survey. Based on survey results, the City developed public relations and "nuts and bolts" strategies to encourage restaurants to adopt Zero Waste activities. These strategies include:

- Customized Zero Waste plans with waste reduction and cost-saving elements.
- Site visits with a focus on product stewardship, consumer action, economic benefits, and education.

Outreach Programs at Fairs/Festivals

Displaying or handing out useful recycling information at community-wide festivals, fair, and other events can also be an avenue for reaching residents in the community.

Reward Systems

Recyclebank

Recyclebank is one of the most prominent customer rewards systems for recycling in the United States and the United Kingdom. It is not a bank, but a company that provides people with redeemable points that can be traded for discounts or products for taking everyday green actions. These actions can include reducing water use, buying "green products," or recycling.

Recyclebank forms partnerships with local solid waste agencies or waste haulers with the objective to increasing recycling rates. It neither does own any recycling trucks nor does process recyclables. Pursuant to these local solid waste programs, residents who register electronically or via the telephone with Recyclebank can report that they have recycled for that week. Residents earn points based on a variety of factors: their individual participation or the total weight of the recyclables set out in their collection area. In return, the residents are rewarded with points that can be redeemed with more than 3000 different vendors, including grocery stores, sports and recreation venues, entertainment, restaurants, and health and beauty establishments.

There are a number of case studies of communities or agencies that have shown significant increases in their community's overall recycling and diversion rates. Some communities, however, have dropped these programs as they have implemented single-stream recycling.

Sources of Public Education Information and Materials

Since the middle 1980s, hundreds, if not thousands, of recycling education programs and materials have been prepared. Much of the material is available to the public for free or at nominal charge. Most state departments of education have collected material or even developed curriculum specific to the population within their particular borders. There are also many state-oriented recycling associations that assist local recycling coordinators with program development. Groups like the National Soft Drink Association publish extensive resource lists which often include specific material and directions for acquisition and price lists.

Local waste haulers and material recovery facility operators often have high-quality educational materials available and may even be willing to sponsor or include recycling coordinators on local educational efforts. There is generally little need to "reinvent the recycling education wheel"; however, great care must be taken to insure that, through efforts to save money, misinformation or inappropriate information is not disseminated to your target audience. Keep in mind that quantity may not be the same as quality when it comes to designing an effective, long-lasting recycling education program for your community.

References

[1] Flavell JW. A tribute to Piaget, society for research in child development newsletter. Chicago, IL: The Society for Research in Child Development; Fall, 1980.

[2] Skinner BF. The behavior of organisms: an experimental analysis. New York, NY: 13 Appleton-Century; 1938.

[3] Geller ES, Lehman GR. Motivating desirable waste management behavior: applications of behavior analysis. J Resour Manage Technol 1986;13:65–7.

[4] Monmouth County. Monmouth County, Solid Waste and Recycling Public Opinion Survey, New Jersey, NJ; 1993.

[5] Beaty EW. Litter got you stumped? Try clean community systems approach: University of Tennessee, MTAS Publications, Tennessee Research and Creative Exchange: Knoxville; 1984.

[6] Rogers Fred, Mister Rogers. Activities for young children about the environment and recycling. Pittsburgh, PA: Family Communications, Inc.; 1990.

[7] Williams, John. Beginning with preschoolers: a recycling education initiative, air and waste management association. In: Proceedings of the annual meeting: Pittsburgh, PA; 1991.

[8] Scott R. The recycle team, WestEd: San Francisco, CA; 1992.

[9] National Agriculture Library. Alternatives to waste disposal. Developed by eEnvironmental Hazards Management Institute, <http://www.nal.usda.gov/ric/ricpubs/waste.html>; 2013. Accessed April 7, 2013.

[10] Connecticut Department of Energy and Environmental Protection, using social media to promote recycling, <http://www.ct.gov/dEep/cwp/view.asp?a=2714&q=487970&deepNav_GID=1645>; 2013. Accessed April 9, 2013.

[11] City of Irvine, California. What is zero waste, <http://www.cityofirvine.us/programs/zero-waste/what-is-zero-waste/>; 2013. Accessed April 11, 2013.

7 Recycling Economics

Market Revenues for Recyclable Commodities

Data Sources

Market price data are a major component in determining the feasibility of any recycling program. This enables the project implementers to develop an effective financial business plan. Consequently, the first step is to gather historical market price data for the recyclable commodities deemed important through a project planning horizon.

There are a variety of reputable, public data sources to gather such information. However, some of the sources listed in the paragraphs below require a magazine subscription or the payment of an additional subscription fee:

- Global Recycling Network—This is a subscription access site, which has been online since 1994. It provides daily spot market prices as well as historical prices for an additional subscription (http://grn.com/prices.htm).
- Platt's Metals Week—Published by McGraw Hill. It contains hundreds of metals' prices across base, minor, light, and precious metals assessed by its editors globally (http://www.platts.com/products/metalsdailyhttp://www.platts.com/products/metalsdaily).
- PPI Pulp and Paper Week—The PPI Pulp and Paper Week is an online subscription that delivers independent reports on the North American pulp and paper industry. The report delivers a weekly news and monthly report covering recovered paper markets as well as other paper grades. It currently incorporates data from the Official Board Markets (often called the "Yellow Sheet") (http://www.risiinfo.com/risi-store/do/product/detail/obm.html).
- RecyclingMarkets.Net—This is also an online subscription-based service, which allows a user to search through its online database for current spot prices as well as historical prices (http://www.recyclingmarkets.net/markets/index.html).
- Resource Recycling—This is a monthly magazine, which also has sister monthly newsletters highlighting e-scrap and plastics. It provides a monthly update of prices, primarily in North America (http://www.resource-recycling.com/).
- Recycler's World—This online database offers historic (quarterly and annual) price reports for different commodities as well as spot market prices on a pay-per-view or on a subscription basis (http://www.recycle.net/exchange/).
- Scrap Magazine—This magazine is published biweekly by the Institute of Scrap Recycling Industries. It contains a series of reports on prices received for scap metals and industry pricing (http://scrap.org/marketreport/index.htm).
- Sound Resource Management—This provides price history graphs (15–20+ years) of monthly prices depending on material for large quantities packed for shipment to end users (F.O.B. processing facilities). Price data is provided for the Northwest United States along with some historical price data for the Northeast United States (http://www.zerowaste.com/pages/Recycling-Markets.htm).

Solid Waste Recycling and Processing. DOI: http://dx.doi.org/10.1016/B978-1-4557-3192-3.00007-5

- The Paper Stock Report—The Paper Stock Report provides a market index of recovered paper prices. It also has an online access subscription with availability of special reports (http://www.recycle.cc/psrpage.htm).
- Waste and Recycling News—This US-based biweekly publication offers a subscription-based online recycling market database (http://www.wasterecyclingnews.com/smp/index.html).

Price Volatility

Price volatility in recycling markets is almost a universal truth across the globe. Being able to manage the ever-fluctuating changes in market prices (Figures 7.1 and 7.2) can either produce success or break the recycling program. Consequently, it is important to help develop strategies in negotiating long-term agreement with brokers or the overall market purchaser that feature price floors or other revenue/risk sharing agreements. It is also important to mitigate these peaks and valleys in these market prices by developing local manufacturing demand for recycled feedstocks.

Most recycling industry observers have opined that prices for most, if not all, recycled materials tend to follow expansions and contractions in the overall world or national economy [2]. There are, however, specific trends in particular industries that move prices for different recycled materials in entirely opposite directions. Experience over the past two decades has shown that communities that collect many different materials may experience less revenue volatility over the course of an economic cycle. Nevertheless, curbside recycling programs that collect a wide variety of materials, such as residential mixed paper, newspapers, cardboard, glass, metals, and plastic bottles, may experience significant and pronounced revenue swings.

Transportation Issues

The distance to the ultimate market is an important variable in assessing the overall economic viability of any recycling program. The cost of transportation can oftentimes exceed the value received by the community for the recyclables during times when market prices are quite low. Free or extremely reduced shipping for recyclable materials is, therefore, absolutely necessary for a recycling program to be economically successful. As such, many communities, for example, in Alaska or in Canada, have worked out arrangements with local barge/air transport companies to backhaul free-of-charge to buyback centers in major cities.

Backhauling

Backhauling is the practice of utilizing shipping space (whether it be in a truck, barge, etc.) to send cargo after a company has already delivered its original goods to their intended destination. For example, if a trucking company delivers a truckload of freight from a major metropolitan area to a grocery store in a remote area, the store might set up an agreement with the trucking company to backhaul, or carry

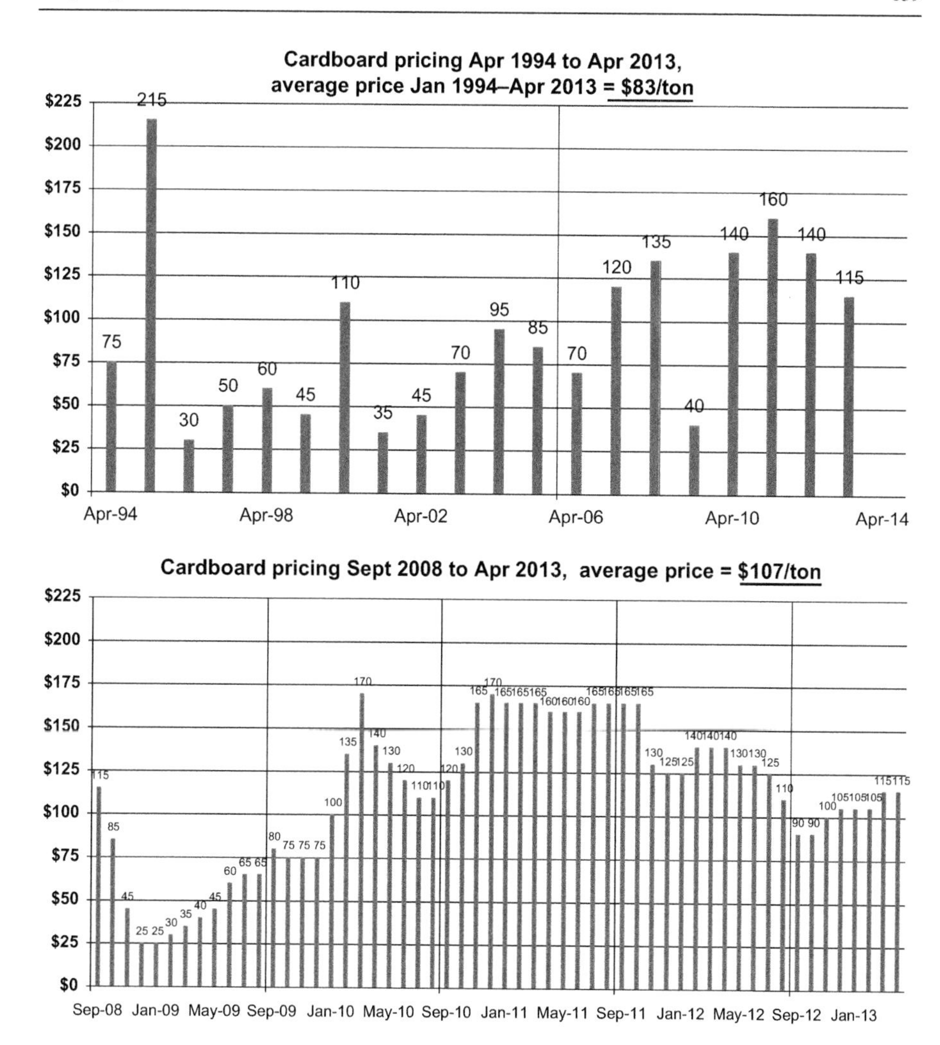

Figure 7.1 Price volatility, 1994–2014. (Victor Horton)

back to the metropolitan area, all of their recyclable cardboard. The practice of back-hauling is becoming more popular as companies are seeking to improve their corporate "green image" or they believe it is part of their corporate social responsibility.

Since shipping costs present the largest stumbling block to many recycling programs, the most important goal of shipping is to be as efficient as possible. In most cases, this means investing in a baler to reduce volume. It also means utilizing the biggest shipping container (to hold the most products) and ship as few times as possible. Of course, space issues may hinder a community from stockpiling goods for as long as they would like, forcing communities "to move" the recyclables constantly. Additionally, market prices may influence when a community ships. If cardboard is

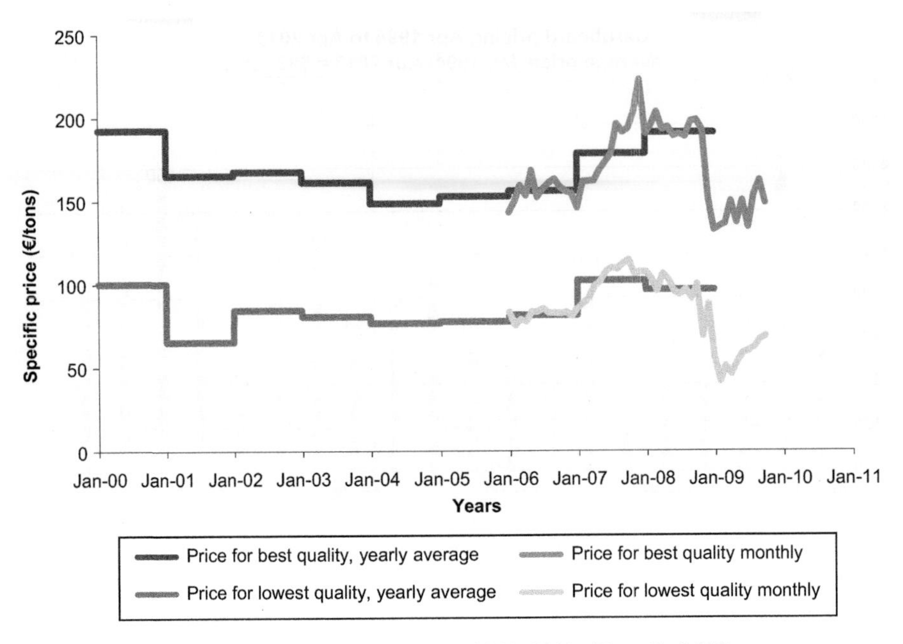

Figure 7.2 Price volatility, recyclable paper, EU 2000–2011. (From Ref. [1])

going for the highest price in 2 years, it oftentimes does not make sense to hold back shipment because the truck is only two-thirds full.

The amount of product able to be shipped at one time varies according to material type and the density and size of bale produced, if any. The standard industry bale size is a 60″(w) × 30″(l) × 48″(h), and usually ranging in weight from 800 to 1200 lb, although this can vary greatly. Typically, most remote communities produce smaller bales than this standard because of the lower volumes of material generated in the communities and the cost for larger baling equipment.

Shipping by Sea

Many marine ferry or shipping companies use 20 and 40 ft shipping containers that they rent to customers. These containers can be delivered directly to the customer by company trucks. Conversely, some customers own their containers with which they arrange delivery and pickup through the ferry system. As mentioned above, using and filling the largest container possible for shipping is always the most economical; so, for the purpose of this analysis, a 40 ft container was used. The maximum tonnage allowed for 40 ft container is 36,000 lb, or 18 tons. Therefore, it is assumed that approximately 36 bales, weighing an average of 1000 lb each, can be delivered in one shipment. Other costs that may be associated with this method of shipping include fuel surcharges and spotting charges, or pickup fees charged by the processors who pick up the goods at the dock to take back to their facilities.

Shipping by Land

A variety of trucking options exist for shipping products on land. Trucks can utilize roll-off containers or trailers. Roll-off boxes are measured in cubic yards, the smallest being 15 CY, although it is more common to see 20, 30, and 40 CY roll-offs being used. The most common trailer sizes range from 48 to 53 ft.

Understanding Cost Analysis

There are a variety of different costs that recycling programs incur and it is important for a recycling coordinator or project manager to understand how cost analysis can help in determining the feasibility of a new or enhanced recycling program. The following sections highlight some of the major cost categories and how they impact a specific cost analysis for a program.

Direct and Indirect Costs

In the world of cost analysis, if a cost can be directly linked to a particular recycling service provided by a community or an agency, then it is usually considered as a "direct cost." These are usually classified into major categories such as labor, equipment, and materials.

To illustrate, a recycling center or composting site will have hourly employees to man these facilities, require utilities such as electricity and fuel, and have public information materials to educate the public about the facility's operations. Further, the typical benefits paid to the employees are also considered a direct cost because they can be tied "directly" to the employees working at the facility.

However, salaried employees are treated differently. For example, if the Solid Waste Director splits his or her time among various operations of the department. Let us say there are four different areas that times are spent on. If the Director spends approximately 25% of his total time between recycling, landfill, collection, and code enforcement, then 25% of his salary would be allocated as a direct cost to the recycling program.

"Indirect costs" are those costs which cannot be easily linked to the particular new program or service. Typically, these are costs where it is usually too time-consuming for the analyst to link to the new service. Table 7.1 provides some typical types of costs, which can be found in most municipal budgets.

Fixed or Variable Costs

Fixed costs are those which remain constant regardless of the level of solid waste or recycling service. For example, salaried employees are typically not eligible for overtime pay in most municipal and private agencies. Consequently, their pay does not increase even if they work additional hours in a week on a project, or if recycling tonnages increase dramatically. Similarly, the depreciation expense for equipment or vehicles does not usually vary as the level of recycling increases.

Table 7.1 Types of Costs

Category	Examples in Agency Budgets
Fixed	Salaried employees Rent Depreciation expense
Variable	Hourly employees Benefits Utilities Fuel Maintenance Landfill tipping charges
Overhead	Human resources, legal, payroll, purchasing, and similar administrative costs for organization Internal phone and mail system Security Management information system Billing services Copier lease
Capital	Debt service for loans and bonds Lease payments

In comparison, variable costs change as the volume or activity level increases. Fuel for machinery or vehicles is a good example. Also, if your charges for processing from an material recovery facility (MRF) are set at per-ton basis, then as you collect more recyclables from a curbside program, your overall costs increase.

Marginal Costs or Savings

Marginal costs are an important concept in evaluating the cost–benefit of making a change in any solid waste or recycling program. Figure 7.3 illustrates a marginal cost evaluation conducted for a community, which was considering a change from manual to automated collection of solid waste. In this case, a number of fixed costs were assumed not to change such as salaried and municipal overhead services. However, variable costs are assumed to change such as fuel, the time it is expected to drive between stops, and the number of potential insurance claims due to worker injuries with the move to automated vehicle collection. Based on this model, the marginal monthly homeowner savings with a change from manual to automated collection was projected to be $2.37, a significant savings.

Time Value of Money

The economic concept of the time value of money is an important one to grasp because it allows the recycling coordinator or analyst to compare the feasibility of

	Number	Unit Cost	Subtotal	Annual	2008 Budget	Variance			
Labor									
ASL Drivers (III)	9.0	$ 47,706.66		$ 429,360					
REL Yard Waste Drivers (II)	4.0	$ 45,435.98		$ 181,744					
Residential Grapple Drivers (III)	6.0	$ 47,706.66		$ 286,240					
Swing Drivers (III)	3.0	$ 47,706.66		$ 143,120					
REL YW Crew (Collector)	8.0	$ 36,485.33		$ 291,883					
Swing Crew	2.0	$ 36,485.33		$ 72,971					
Subtotal	32.0			**$ 1,405,317**	2,539,943	(1,134,626)			
Equipment									
ASL Vehicles (front line)	9.0	$ 239,000.00	$ 2,151,000	$ 430,200					
ASL Vehicles (spares)	3.0	$ 239,000.00	$ 717,000	$ 143,400					
REL YW Vehicles (front line)	4.0	$ 197,146.00	$ 788,584	$ 112,655					
REL YW Vehicles (spares)	2.0	$ 197,146.00	$ 394,292	$ 56,327					
Grapple Vehicles	7.0	$ 106,349.00	$ 744,443	$ 106,349					
ASL Containers	43000	$ 45.00	$ 1,935,000	$ 193,500					
ASL Containers (spares)	2150	$ 45.00	$ 96,750	$ 13,821					
Subtotal			$ 6,827,069	**$ 1,056,253**	459,548	596,705			
Operating Costs									
Maintenance									
ASL	12.0	$ 2,744.80	$ 32,938	$ 395,251					
REL	6.0	$ 1,700.75	$ 10,205	$ 122,454					
Grapple	7.0	$ 1,220.00	$ 8,540	$ 102,480					
Subtotal				**$ 620,185**	654,622	(34,437)			
Fuel	25.0	$ 1,700.10	$ 42,503	**$ 510,031**	530,432	(20,401)			
Disposal	54519.00	$ 29.33	$ 1,599,042	**$ 1,599,042**	1,573,146	25,896			
Administration Allocations									
Residential				$ 1,426,204					
Grapple / Appliance				$ 253,421					
Administration allocation adjustment as provided by City				$ 167,553					
Potential Reduction in Self-Insurance Fund			-30%	$ (97,394)					
Subtotal				**$ 1,749,784**	2403783.44	(653,999)			
							current		
Total				**$ 6,940,612**	**$ 8,161,474**	**$ (1,220,862)**	$ 15.75	$ 2.30	$ 1,186,388
Fully Loaded Administrative Costs (FY2009)				$ -					
Cost Per Household							Potential Savings		
Annual				$ 161.41	$ 189.80		savings	monthly	annual
Estimated Monthly Cost per Household:				$ 13.45	$ 15.82		$ (2.37)	$ (101,739)	$ (1,220,862)

Figure 7.3 Example of marginal cost analysis comparing collection program changes.

different alternative solid waste programs. Essentially, it is based on the simple rule: a dollar invested tomorrow is worth less than a dollar received today.

To illustrate this concept, let us look at a typical example for a recycling project. If the income stream today from an MRF project is $500,000 a year, the agency could earn some interest on this cash income. At 5% interest, this cash income would earn about $25,000 a year or approximately $70 a day. So, at a 5% interest rate, the opportunity cost of receiving your $500,000 tomorrow rather than today is $70. As the interest rate increases, the agency gains more by deciding to take the cash flow today.

The time value of money concept plays out in many purchasing decisions such as making an investment by purchasing new vehicles, baling equipment, a tub grinder, or public informational materials. Typically, the upfront investments for these expenditures are paid in today's dollars, while the cash flows from the potential savings are paid in future dollars which are worth less than today's dollars because of the time value of money.

Capital and Operating Costs

Capital costs are the big ticket expenditures for a solid waste and recycling program. These include items such as vehicles, equipment, and buildings, which have an expected life span of several years. In typical cash accounting systems, these expenditures will be recorded as the full expense for the first year and zero dollars for the remaining life of the item. Under accrual or full cost accounting, depreciation comes into play, which is a method of allocating these purchase costs over the useful life of the asset. Depreciation takes into account three different variables: the initial purchase price of the asset, its expected useful life, and an estimated salvage value at the end of its life. There are several different depreciation methods depending on potential tax code implications. However, the simplest is straight line depreciation which follows the following formula:

$$\text{Depreciation} = \frac{\text{Cost-salvage value}}{\text{Life in number of periods}}$$

In comparison, operating costs are the normal reoccurring costs that are used or consumed over a short period of time, typically less than 1 year. These include budget items such as wages and benefits, rent and lease payments, fuel, maintenance costs, and interest or debt service payments.

Determining Economic Feasibility

All of the economic concepts discussed in the previous sections can be used by the Recycling Coordinator or Project Manager to help determine the economic feasibility of a particular recycling project. Most analysts utilize individual spreadsheets or linked spreadsheets to automate the feasibility process. These are briefly described using a few examples from the author's consulting experience.

Simple Mathematical Analysis

Financial and economic analysis for project feasibility encompasses a range of tools from simple mathematical calculations to those using computer applications. These will be briefly discussed with a number of specific illustrations.

Breakeven Analysis

Breakeven analysis is defined as the point in a solid waste or recycling project when its total cost equals the money saved in things such as waste collection, transportation, and ultimate disposal costs. That is, the point in the project when there is zero loss. The usual formula for calculating the breakeven point is as follows:

$$\frac{\text{Fixed costs}}{\text{Revenue per item} - \text{cost per item}}$$

Typically, most recycling programs are designed to produce savings from solid waste disposal by diverting recyclables from the community's waste disposal containers. For example, consider the situation where a community implements a curbside recyclables collection program. Its collection contract with a private waste hauler is $1,000,000 per year. The materials collected go to a regional MRF facility with a tipping fee of $50 per ton. This figure includes all disposal costs of the residue. The community receives an average revenue of \$25 per ton. The breakeven level for the community can be calculated as follows:

$$\text{Breakeven} = \frac{\$1{,}000{,}000}{\$50+\$5} = \frac{\$1{,}000{,}000}{\$55} = 18{,}181\,\text{tons}$$

In this specific case, the recycling program will reduce the cost of the community's solid waste disposal if more than 18,181 tons are collected.

Payback Period

While breakeven analysis takes into account the number of unit that must be recovered to return an investment, the payback period tells how soon this investment will be paid back. This is extremely important where there is a significant capital investment for a project and the cash flow revenue stream is extended over a long period of time.

There are numerous examples in recycling that utilize this concept. For example, a regional mall, which generates a significant amount of food waste, decides to evaluate the feasibility of installing a food-waste composting unit. Their current cost of contracting with a private waste hauler is \$3400 a month in hauling and disposal charges. The monthly cost of the composting unit is quoted by the manufacturer at \$2100 with maintenance (wood chips and microorganism solution), electrical, and additional water use is \$140. The payback period is less than 2 months.

Pro forma Economic Modeling

Any Pro forma model must be developed from a grasp of the dynamics of the market influencing the life cycle of the project. While software programs can be great tools, the programs are only as good as the assumptions that go into the program. Another common problem with most Pro forma models is the "one size fits all" syndrome. Every project is unique and the design of the Pro forma financial model should reflect these differences. To accommodate the various types of business models needed to analyze the feasibility of recycling projects, the author has developed a variety of different types of Pro forma models that has allowed him to tailor the financial statements to the specific project. This has provided clients the models with maximum flexibility to model multiple scenarios of facility size, biogas production/cogeneration, and site locations.

Figure 7.4 illustrates the general configuration of a Pro forma model. As shown, a model is crafted to develop very flexible financial assessment, which incorporated major project feasibility factors to help answer many "what if" questions.

Data Inputs

Using the cost assumptions and critical project assumptions (Figure 7.5), a multiyear (typically 5- to 20-year) projection of projected revenues, operating expenses, and

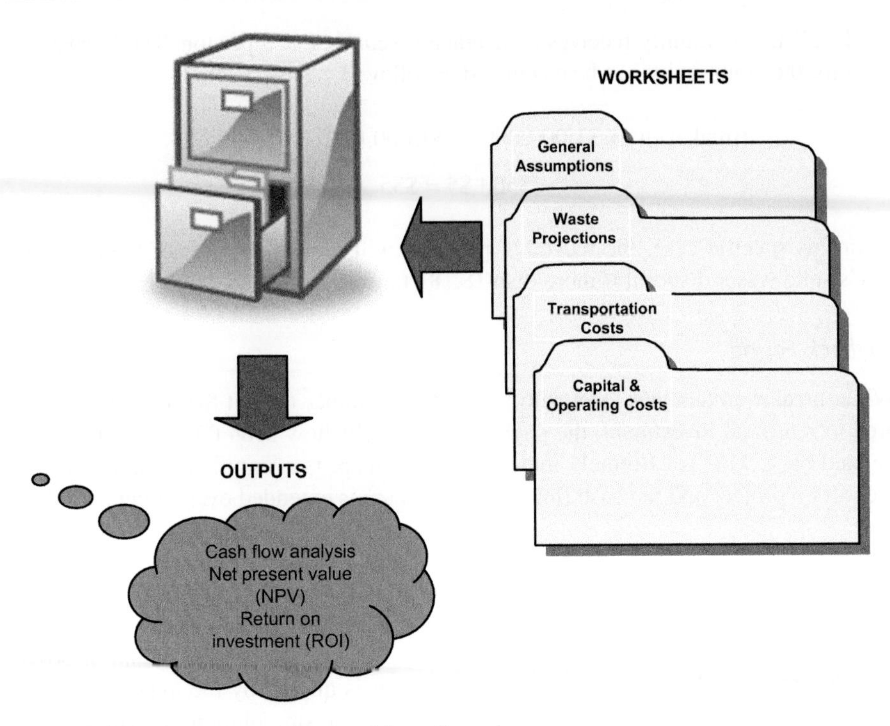

Figure 7.4 Pro forma spreadsheet model configuration.

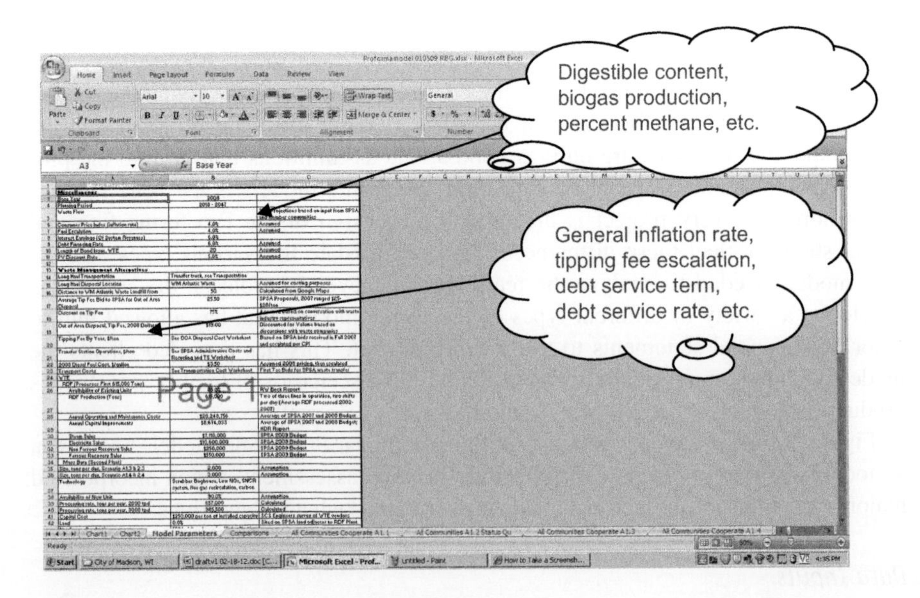

Figure 7.5 Typical pro forma model assumptions.

debt services will be developed using Microsoft Excel. Assumptions are usually based on working knowledge of the solid waste industry, recently reported case history, and actual ranges in capital and operating costs for similarly sized recycling facilities.

Waste Received

Operating results will be based on the projected quantities of municipal waste and/or recyclables generated by the solid waste agency and possibly adjacent communities. The waste flows spreadsheet will help estimate the projected flows of solid waste and/or recyclables for the proposed recycling facility. These data can incorporate results from a community's pilot program as well as benchmarking available from similarly sized municipalities.

Capital Costs and Operating Expenses

The historical capital and operating costs for the different recycling systems under examination are the first step in estimating the costs of a proposed solid waste or recycling program. In cases where there are limited capital and operating data, it is commonplace to collect data on similar recycling programs or facilities from both the literature and through formal manufacturer's quotes.

The capital costs should include all predevelopment and construction costs. Operating costs will typically include labor, maintenance, materials, testing, insurance, potable water, waste services, overheads, and training costs, as well as potential costs for taking residuals (contaminants) to waste disposal sites, including any transportation costs or required tipping fees at these facilities (Figure 7.6).

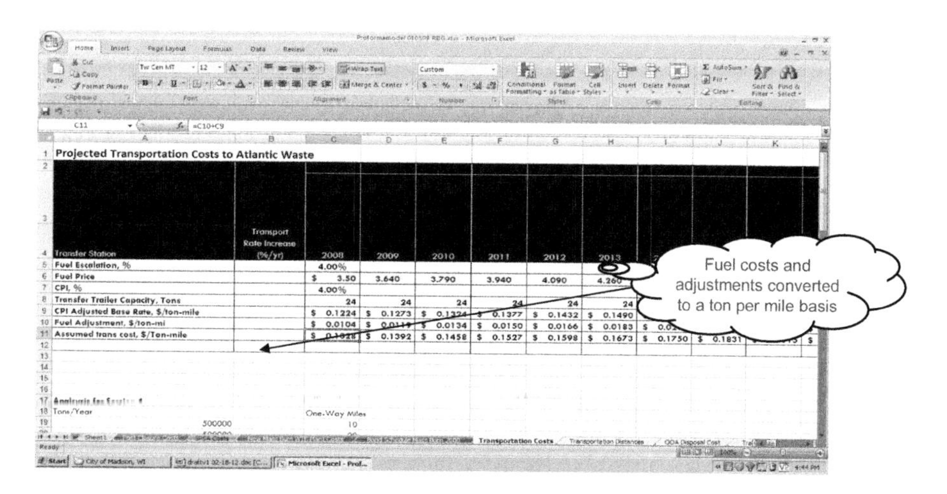

Figure 7.6 Transportation cost spreadsheet.

Operating Revenues

The end game of an economic feasibility assessment study is to prepare and estimate the cash flows of the project over its useful life and determine at what rate they should be discounted. After putting the projected revenue streams into the model as a starting point, the model must include assumptions about the future, including future energy (electricity, biogas, or steam) revenues, tipping fees, and revenues from the sale of marketable products. All are elements that must be estimated to build the revenue side of the Pro forma model.

Revenues for recycling projects can come from any combination of the following sources:

- Recyclables
- Energy (gas, heat, electricity)
- Secondary products (compost, water, liquid fertilizer, feedstock for City compost)
- Renewable energy tax credits, carbon offset credits, grants, etc.

For example, an anaerobic digester facility sited at a waste treatment facility could offset the community's electricity demand, in which case the revenue for electricity would be based on the retail price of electricity which may be 25–40% higher than wholesale rates received from the utility. Also, since the facility can be expected to produce a large quantity of biogas, there may be a possible option to upgrade the biogas to natural gas so that it could be injected into pipelines or can be used to produce compressed natural gas (CNG) for use by municipal fleet vehicles.

Financing Options/Alternatives

In order to complete the feasibility assessment of the proposed project, the analyst must evaluate current potential options and alternatives for financing the project which could include the following possibilities:

- Loans
- Municipal bonds
- Governmental grants
- Private vendor financing.

These findings of this assessment can help prepare different economic scenarios of the Pro forma model of the proposed project.

Life Cycle Analysis

To complete the valuation of the project, the reversion value or future project value must be calculated. A Pro forma model should include two valuation pieces. The first is the net present value of the net cash flows during the period of projection. The second piece is the value of the asset at the end of that projection period, the so-called reversion value. This value should represent the capitalized value of the normalized net operating income of the project or the projected market revenue for the project less the normalized expenses, including improvements and capital reserves. A capitalization rate can then be applied to calculate the reversion value. This value must then be discounted to the net present value. These can then enable the feasibility engineer to provide a life cycle financial analysis of the project (Figure 7.7).

Presenting All the Numbers

All of the modeling and spreadsheets are meaningless if the analyst is unable to communicate the results to the decision-makers.

Cost per ton

Cost per ton is a common economic benchmark reported in solid waste and recycling. It is easy to calculate, and it should include all costs and revenues to ensure that it accurately represents a number that can be compared to typical landfill and waste-to-energy costs. As given in Table 7.2, the data attempts to account for transportation and processing costs of recyclables to the secondary market, as well as anticipated revenues based on an assumption of projected delivery amounts.

Cost per Household

Cost per household is a popular economic benchmark for recycling coordinators as well as the general public. Oftentimes, this benchmark is used in annual advertisements or public notices to provide the public with information on the cost of recycling and collection services (Figure 7.7).

Table 7.3 illustrates a typical calculation of the cost per average household or customer for a new recycling program or improved service.

BOARD OF COUNTY COMMISSIONERS
INDIAN RIVER COUNTY, FL
SOLID WASTE FULL COST ACCOUNTING PUBLIC DISCLOSURE NOTICE

In accordance with the requirements of F.S. 403.7049 and Section 62-708.500, FAC, the Board of County Commissioners is advising all users of solid waste management services in Indian River County of the information below concerning the full cost of solid waste management.

Full Cost of Solid Waste Management Services
Related to Collection, Disposal and Recycling Services
Provided by Indian River County Solid Waste Disposal District
Fiscal Year 2011-12

TOTAL COSTS ALLOCATED:

Function	Total Amount	Amount Allocated to Residential (1)	Amount Allocated to Non-Residential
Disposal	$5,258,775	$2,740,028	$2,518,746
Convenience Center & Recycling	5,400,229	3,554,186	1,846,043
Total	$10,659,004	$6,294,214	$4,364,789

FULL ANNUAL ASSESSMENT COST:

	Average Annual Cost Per Household Single Family	Average Annual Cost Per Household Multi Family (2)	Per Ton
Disposal:			
Residential	$29.53	$22.15	
Non-Residential			$16.96
Convenience Center & Recycling:			
Residential	$38.31	$28.73	
Non-Residential			$12.43
Total Costs	$67.84	$50.88	$29.39

(1) Residential category includes single family units, multi family units, mobile homes and travel trailer sites.
(2) Includes multi-family, mobile home and travel trailer dwelling units.
(3) Additional costs for collection are associated with several of the individual jurisdictions.

Figure 7.7 Notice for cost of collection and recycling service. (Indian River County, Florida)

Table 7.2 Example of Cost per ton Analysis

Capture Rate (%)	Total Recyclable Tonnage	Plastics (tons)	Glass (tons)	Steel Cans (tons)	Aluminum (tons)	Trips per Year if Baled (1)	Transportation Cost per Year to Haul to Whitehorse (2) ($)	Estimated Disposal Cost Charged by Raven ($200/ton Processing Fee for Mixed ($)	Estimated Revenue from Aluminum (5) ($)	Total Yearly Cost to haul to Whitehorse ($)	Net Cost per ton of Recyclables ($)
10	11	3	4	2	2	1	590	1748	821	1518	144.19
20	21	5	8	4	4	1	590	3497	1641	2445	116.16
30	32	8	12	6	5	2	1180	5245	2462	3963	125.50
40	42	11	16	8	7	2	5739	6993	3283	9450	224.45
50	53	13	21	10	9	3	7174	8742	4103	11,812	224.45
60	63	16	25	12	11	3	8609	10,490	4924	14,175	224.45
70	74	19	29	14	12	4	10,044	12,238	5744	16,537	224.45
80	84	21	33	16	14	4	11,478	13,987	6565	18,900	224.45
90	95	24	37	18	16	5	12,913	15,735	7386	21,262	224.45
100	105	27	41	20	18	6	14,313	17,432	8206	23,539	224.18

Table 7.3 Example of use of Cost per Residence or Customer

Revenue/Cost Items	Revenues/Cost of Service	
	Annual ($)	**Monthly (per Residence) ($)**
Estimated revenues		
Recyclables	823,500	2.15
Subtotal	823,500	
Estimated operating expenses		
Labor	0	2.08
Transportation	228,880	
Processing	503,360	
Public education	65,000	
Subtotal	797,240	
Estimated capital costs		
Vehicles	0	0.59
Carts	226,633	
Subtotal	226,633	
Total net cost of service	200,373	0.52

Diversion Rate

The diversion rate is a measure of how much material is diverted from disposal to recycling in a community program. These rates are usually calculated from agency or hauler records using the following formula:

$$\frac{\text{Tons of recyclables collected}}{(\text{Tons of recycling collected} + \text{Tons of solid waste collected})} \times 100$$

From purely a cost perspective, the more important benchmark is the percentage of recyclable materials diverted from the community's disposal location.

Participation Rate

Oftentimes, the participation rate is the most quoted benchmark, but, in the author's opinion, it is the least reliable benchmark. While it measures how many households are participating in a curbside recycling program, it tells very little how much they are recycling. For example, a homeowner could only set out plastic bottles and aluminum cans, not all the materials desired in the recycling program, and be counted as participating in the overall program.

Financing Programs

Grants

Many states in recent years have funded local community recycling programs through program grants. Such funds have been used by communities to initiate the planning of recycling programs or they paid for the initial equipment acquisition for the community program. Other states have allowed grant funds to pay all or a major portion of the capital costs of facilities such as MRFs or the short-term operating costs of a curbside recycling program. Advanced disposal fees (ADFs) have been used by a number of states to fund grant programs which have then been used to finance various local government recycling programs. An ADF is a fee imposed on a product, such as tires, motor oil, white goods, batteries, and containers, which is intended to reflect its cost of disposal. By levying a fee on these consumer products, proponents argue that the consumer will then have an economic incentive to buy products which are more "environmentally friendly" and enhance their demand. However, research conducted by Arthur D. Little, Inc., has suggested that the low fees imposed in states which have enacted ADFs has little, if any, impact on reduction in demand for these products. Further, experience of states which have enacted ADF legislation has shown that ADFs are oftentimes difficult and costly to implement. Nevertheless, ADFs have been popular with many state legislatures since ADFs can provide a substantial funding source.

Bonds

Some recycling projects are capital intensive. Most communities do not have the available capital or tax base to "pay-as-you-go" during construction or procurement, so alternate financing mechanisms must be used. To the extent that a source of money is available to offset a portion of the construction cost, the amount of financing required can be correspondingly reduced.

The two primary expenses associated with financing are the interest rate on bonds or the expected internal rate of return for private funds and the cost of capitalized interest. The amount of capitalized interest is usually 35–40% of the bond size, and this will vary according to interest rates and the length of construction. The interest rate or rate of return is influenced by: (1) the bond market at the time of sale and whether bonds bear interest at fixed or variable rates; (2) whether the interest is tax-exempted, and, if so, the extent of the federal tax-exemption; and (3) the security structure of the bond and credit quality of the obligations.

There are three financing options generally available for recycling projects: (1) grants, (2) bond sales, and (3) private funds. In addition, a combination of these options may be utilized. The following discussion focuses on bond sales and private funds because grant funds for construction of these projects are limited in most states. Bond structures most commonly used with respect to recycling facilities include: (1) general obligation bonds and (2) revenue bonds. The most significant distinction between these alternatives is the nature of the security structure; in particular, the level of obligation assumed by the governmental entity.

General Obligation Bonds

General obligation bonds would pledge the full faith and credit of a community as security for payment of the bonds. The full faith and credit pledge includes the general revenue of the community which may include property tax, sales tax, license fees, income tax, and other charges as payment for the bond principal and interest. The bond interest rate reflects the community's credit worthiness with less emphasis on project economics. General obligation bonds are generally tax-exempt. However, if the bonds meet the criteria to qualify as private activity bonds, the tax-exemption is limited to taxes on the gross income of the bondholder. In addition, the bonds must meet additional criteria in order to qualify even for this limited exemption. In determining whether to use general obligation bonds, the following should be taken into consideration:

1. Depending on the credit rating of the community, the interest expense may be lower than the other alternatives discussed.
2. The community will assume maximum financial exposure which may lead to downgrading of its credit.
3. The obligations or bonds will impact the community's debt limit and consequently reduce the community's borrowing capacity.
4. A public vote will be required to approve sale of the bonds.

This type of financing has not been used by many communities to finance construction of its public works facilities and it is unlikely that general obligation bonds would be issued for a recycling project.

Revenue Bonds

Revenue bonds pledge the revenues available from the project as security for the bonds. The project revenues may include solid waste disposal fees, revenues from the sale of recovered materials, any available insurance proceeds, and damage payments from contractors, if applicable. Most revenue bonds are limited obligations, and as such, the community would not be obligated to raise taxes to pay debt service. However, the community will generally be required to covenant that rates charged for solid waste collection and disposal will generate revenues sufficient to cover all project costs including bond principal and interest payments. This "rate covenant" enhances the project's security and may help to lower interest rates. Another security feature is establishment of various bond reserve funds to provide a degree of protection against revenue shortfalls or unexpected operation and maintenance expenses.

Since project revenues are the security for revenue bonds, the bond interest rate will reflect how project risks are handled and the bond buyer's exposure to project risks. The bond holder's risk exposure is typically judged by bond-rating agencies such as Standard & Poor's or Moody's.

Another type of revenue bond financing is often termed "Conduit Financing," although such financing is severely limited. Conduit financing along with private ownership can be used in recycling projects when the tax benefits available to private corporations (as owners) offset project costs and, consequently, lower disposal fees. A private investor is considered eligible for certain tax benefits by contributing only a portion of the project costs from private funds with the balance of project funding supplied by tax-exempt, private activity bonds (PABs) issued by a community on behalf of the private entity. The portion of the project costs contributed by the private owner is called the equity contribution.

The equity contribution will usually be provided during the construction period on a pro rata basis with other construction expenses. The private corporation may also provide a corporate guarantee for payment of the bonds, subject to certain conditions, as an additional security feature for the PABs. If an unconditional corporate guarantee is provided, the rating agencies will focus on the credit worthiness of the corporation as well as the project economics. However, in order to obtain an unconditional corporate guarantee, the corporation must believe that the project is an excellent risk.

As discussed above, in an equity participation arrangement, the contractor is also the project owner. Generally, the contractor is a limited subsidiary of a large parent corporation. The parent corporation must have a tax benefit appetite in order to take maximum advantage of available tax benefits. Because not all contractors can effectively use the available tax benefits, the transaction may be structured as a sale/lease back with a third-party owner with a large tax benefit.

The bond interest rate will also be influenced by the tax-exempt status of the bonds. PABs are subject to many limitations which are not applicable to other bonds (which are not PABs), including restrictions on the type of facility financing and restrictions on the use of proceeds to pay issuance costs. PABs may not be refunded in advance, even if the criteria for tax-exemption are met. Interest on the bonds will be exempted from the gross income of the bondholder but will remain subject to the alternative minimum tax on individual and corporations. As a result, PABs will bear higher interest rates than would non-PABs with comparable security features.

Private Funds

In addition to the revenue bond structure described above, there have been proposals for 100% private financing of recycling projects. Under this option, the private firm would raise the capital, design the recycling project, procure the equipment and/or construct the system, and operate the program. There are many options to this type of private alternative in terms of procurement, operations, and degree of ownership. However, till now sole private financing has not been evidenced as a preferred method of private ownership and is not considered a viable option for most communities. There are several reasons why 100% private financing has not been implemented:

1. The capital intensive nature of the project requires substantial corporate commitment. Investment of 100% of the project cost "ties up" capital which could be used in other projects.
2. Tax benefits have been reduced in recent years and may be received for an investment of only a small portion of the project cost.
3. Financial analysis of the corporation's costs of money and required rate of return on 100% rather than a small portion of capital investment typically results in a higher operating fee requirement.

Bank Loans and Anticipation Notes

Many local governments have utilized bank loans for short-term project financing (rolling stock, site purchases, vehicles, etc.) since they do not have available capital out of their general fund to "pay-as-you-go" during the initial procurement of equipment or construction of the recycling facility. Bank loans are used until long-term bond debt can be issued under more favorable market conditions. These are termed Bond or Tax Anticipation Notes, BANs and help communities to stabilize their cash flow. In some areas of the country, however, the interest rate on such loans can compare favorably with municipal bond rates, particularly since bond issuance costs are saved or deferred until a bond issue refunds the loan at a later date as the notes expire. Further, many local banks have long-term relationships within a community, thereby having a greater understanding of the local need and support for a project. This is often reflected in reducing the need for restrictive project covenants and complicated financial negotiations with parties distant from the community.

Leasing

Equipment leasing is prevalent in the solid waste industry, particularly with the acquisition of collection vehicles. Most often, an equipment manufacturer secures a leasing company purchases and holds title to the equipment during a specified period, commonly the usual length of depreciation of the asset (5–7 years). During this term, the community or private operator pays the leasing company a lease payment for using this equipment. At the end of the lease agreement, the lessee can purchase the equipment at a value specified in the lease agreement. This type of lease arrangement is often favorable to the community since it defers the long-term capital funding of expensive, but depreciable, equipment.

Another form of equipment or facility leasing is often termed "leveraged leasing." Under this type of leasing system, with the private entity (lessor) providing a portion of the capital needed to purchase the asset. The community then finances the balance of remaining capital needs with a financial intermediary who acquires the tax advantages of ownership. This allows this third party the right to depreciate the asset and perhaps be able to receive an investment tax credit, if one is available under federal law and certain state tax codes for recycling facilities and equipment. It is these tax advantages that allow the third party to receive a higher after-tax return on its investment which may be reflected in the lower interest on the cost of money for the community compared to other financing methods.

Customer Billing Approaches

There have been a great number of approaches utilized by solid waste agencies to help pay for the cost of providing enhanced solid waste recycling programs. These are addressed in the paragraphs that follow.

Property Taxes

Although many communities have established solid waste systems founded on an enterprise-based accounting program, property taxes continue to provide the basis of funding in many communities throughout the world. Funding for solid waste services, including collection, transport, recycling, waste reduction, and recycling services can be usually found in the general revenue accounts of the community. This usually means that no separate billing or collection system is required thus making property taxes easy to administer. Additionally, many residents prefer this funding option because local property taxes are deductible on federal and many state tax returns.

There are, however, several major disadvantages to this type of funding. Unfortunately, this type of funding program does not accurately account for the "true" or "full" cost of providing solid waste services since many related municipal services (e.g., fleet maintenance, insurance, legal, accounting, etc.) are commonly left unaccounted. Thus, the cost of municipal solid waste collection and disposal programs under these circumstances have been traditionally considered to be "free" services by many residents and businesses, thereby hindering the development of effective recycling programs. Some have argued that only when residents and businesses are charged, the real costs of providing solid waste services will have a real chance of success. Furthermore, local property tax caps greatly restrict the overall level of funding for recycling programs since these programs must compete for limited public funds for other essential public services such as police, fire, education, and health.

Assessment Systems

A number of states currently allow local governments to fund the cost of their solid waste system operations through the imposition of special solid waste assessments on users. Typically, these assessments are placed on the non-ad valorem portion of the property tax bill on residences and/or commercial businesses within a specified service area, commonly called a municipal service benefit unit (MSBU). Billing for refuse service through a combined tax bill mechanism is less costly than if a separate tax bill would have to be sent out. Failure to pay these annual assessments is treated by many government jurisdictions as that of failure to pay property taxes. In some political jurisdictions, a lien is either placed on the property, which must be paid when the property is sold, or the lien is sold as a tax lien by the government entity to third party. Very often, governments are hesitant to impose such conditions because individual homeowners who pay little or no property taxes may be forced out of their homes for failure to pay these non-ad valorem assessments. Some communities have addressed this problem by providing specific exemptions from paying these assessments by allowing senior citizens to have low-income discounts. This type of special assessment has gained favor in communities with limited waste stream control.

The assessment is established by the government entity to cover the cost of the service, in this case, solid waste collection, recycling, and/or disposal service. Generally, a flat fee is levied for the service regardless of the service level provided to all residential units and/or commercial businesses within the benefit unit. The benefit unit may consist of an entire franchise collection area or the entire governmental jurisdiction (e.g., incorporated or unincorporated areas). However, having separate benefit units for each franchise area could allow for different rates between different franchise areas.

However, assessment systems have their own disadvantages. Commonly, a separate tax roll must be prepared which will include all properties within the benefit unit. Preparation of this tax roll may be somewhat costly, especially if there is poor cooperation between the government entity imposing the special assessment and the agency which formally issues the tax bill. Also, such assessment fee systems are often unpopular with local citizenry unless an effective public information program is developed which explains the need for the assessment program.

A good example of a special assessment program used to fund a local recycling program can be found in Charlotte County, Florida. In 1992, the County established a special assessment program in the unincorporated area of the County for billing the fee charged for disposal of solid waste, including the cost of the curbside recycling program, on the annual tax bill. Under this program, all residential properties of four units or less per building (improved property) are assessed for solid waste collection and disposal services, with a credit against this special assessment given to those residential properties which subscribe to franchise hauler services and pay the special assessment through their franchise hauler billing. Exemptions from paying the special assessment are also provided to properties with a permit to haul solid waste which is generated on their property to a transfer station or drop-off facility. The special assessment is collected through the office of the Charlotte County Tax Collector.

Sales or Municipal Taxes

Sales taxes can provide significant funding for recycling programs, especially in areas which are heavily populated or have high tourism and recreational activities. Some states, like Florida and Georgia, allow communities to levy local option sales taxes on gasoline or other consumer goods and apply these sales tax revenues to the construction and/or operation of solid waste facilities. Generally, however, these levies are often inadequate for larger projects. Many states also allow local communities, primarily municipalities, to levy taxes on utility services (telephone, electricity, cable TV, gas, water, etc.), utilized by residents in their community. By state statute, the funds generated by these levies can then be used by the community for any lawful purpose. Again, as in the case of property taxes, these revenues are often used for essential public services which recycling services must compete against.

Variable Rate Systems

Pay-as-you-throw, commonly called PAYT, is a solid waste rate strategy that charges households a higher bill for putting out more trash for collection [3]. As of now, more than 7000 (25%) communities in the United States agree and use some form of PAYT, including 30% of the largest cities. For a number of years, the EPA in the United States has promoted PAYT through webinars and dedicated web sites (www.paytnow.org) with PAYT resources available to communities across the globe.

PAYT (also called variable rates, volume based rates, user pay, and other similar names) provides a different way to bill for solid waste collection service. Instead of paying a fixed bill for unlimited collection, these systems require households to pay more if they put out more garbage—usually measured either by the can or by the bag of garbage. Paying by volume provides households with an incentive to recycle more and reduce disposal.

Communities have been implementing PAYT trash rate incentives in earnest since the late 1980s. The programs can provide a cost-effective method of reducing landfill disposal, increasing recycling, and improving equity, among other effects. Experience in these 7000 communities, which are distributed all across North America, shows that these systems work very well in a variety of situations:

- Private haulers, multiple haulers, or municipal collection
- Manual or automated collection trucks
- Cart or other types of containers
- Urban, suburban, small/rural, and isolated communities
- Set up by ordinance, by contract, or municipally run.

How PAYT Works

The most common types of PAYT systems are [4]:

- Variable can or subscribed can programs ask households to sign up for a specific number of containers (or size of wheelie container) as their usual garbage service level and get a bill that is higher for bigger disposal volumes. This is a common choice in areas with fully automated trucks using lifting arms.
- Bag or sticker/tag programs require households to buy specially marked bags for trash; the bag price includes the cost of collection and disposal. Bags are usually sold at convenience and grocery stores in addition to municipal outlets. Other programs require households to buy special tags or stickers to place on bags or cans; pricing is similar to the bag option.
- A hybrid program uses the basic system—households keep paying a bill they have always paid (to the city or hauler), but instead of covering the cost of "all" or unlimited amounts of trash, it only covers 30 or 60 gallons. To get more service, special bags or stickers must be purchased (as above). This system combines existing programs and new incentives, and minimized billing and collection changes. Some rural communities have drop-off programs, where customers pay by the bag or weight at transfer stations using fees, bags, stickers, or prepaid punch cards. Some haulers also offer PAYT as an option, or customers may choose unlimited service for a fixed fee.

Methods for Implementing PAYT

There are three key methods for implementing PAYT, and with political will behind it, the speed with which it can be implemented is rather astonishing at the community level. Of course, the authorities and rules vary in different states, but the following summarizes the key options.

- Ordinance: a community passes an ordinance specifying that any hauler operating in the community (or county) must use a PAYT system for residences. While rates cannot be specified, rate *structures* can. The best PAYT ordinances include a requirement for haulers to provide curbside recycling service (with the ordinance specifying minimum recycling materials, containers, and frequency); PAYT rates that cover the costs of both trash and embedded recycling costs; small trash container option; PAYT rates that have significant price differentials between service levels (the www.paytnow.org web site has detailed advice on these issues); and reporting/auditing requirements. This is a fairly easy option because it does not "take" business from haulers, rather offers a business opportunity to collectors (recycling for all, not just some or none), maintains competition, and can be implemented quickly with minimal hassle for the community.

- Contract, districting, franchising: an RFP or other service document can be issued, soliciting service providers interested in providing service in a geographic area, with PAYT (and the types of options above) as part of the contract. This provides communities with more control and possibly lowers rates if economies of scale are not in effect in a community, but can be more politically tricky because it "takes" business from some haulers. Of course, some advance notification and other requirements are commonly involved in invoking this option. However, this can be a revenue source for communities, either through hauler/franchising/other fees, or, if a city opts to bill for the service, they may potentially charge rates that cover both the contractor and additional recycling programs and efforts.
- Municipalization: if a community already provides service directly (or opts to go in that direction), implementing PAYT is quite straightforward.

References

[1] Eurostat. 2013. <http://epp.eurostat.ec.europa.eu/portal/page/portal/environment/data/main_tables, European Commission, Luxembourg>.

[2] Morris J, DeFeo W. Practical recycling economics—making the numbers work for your program. New Jersey, NJ: Department of Environmental Protection, Division of Solid and Hazardous Waste: Trenton, New Jersey; 1999.

[3] Skumatz LA, Freeman J. PAYT: 2006 update, for US EPA and SERA. Skumatz Economic Research Associates, Superior, CO.

[4] Skumatz LA, Rogoff M. Pay-as-you-throw now: increase recycling and decrease greenhouse gases quickly, fairly and cost effectively. APWA Reporter: Kansas City, Missouri; March 2010.

8 Institutional Issues

Procurement

In general, procurement encompasses the acquisition by a community of equipment and operational services. Procurement can be discussed in terms of approach and procedure. The procurement approach dictates the manner by which engineering, design, construction, start-up, and operation are acquired and how responsibilities will be assigned between the public and private sectors. The procurement procedure determines the method by which such services are acquired and the legal guidelines which must be followed. The following discussion identifies the approaches and procedures generally available, the relationship to various risk assignments, and the analysis of the appropriateness of the available procurement alternatives.

Procurement Approaches

There are essentially three approaches used in recycling procurement: (i) design, bid, build (D–B–B approach); (ii) design, build (D–B) approach; and (iii) design, build, operate (D–B–O) approach.

Design, Bid, Build (D–B–B) Approach

The D–B–B approach is the traditional and most widely used approach for procuring solid waste and recycling projects. Typically, a professional engineering firm is retained by a community to participate in the planning and design of the project. The engineer, acting as an agent for the community, prepares equipment and system specifications to be let for public bidding, and is responsible for certain design elements of the project. Following bid evaluation, the same, or a different; engineer is retained by the community to monitor construction of the project in order to ensure the use of proper materials, supplies, equipment, and so on. Upon completion of the construction, the engineer assists in plant start-up and testing, and may be required to prepare operating manuals for the project. Once the project has passed acceptance testing, operational responsibility becomes that of the community which might either operate the project itself or contract for its operation with a private firm. This process usually requires several different contracts between the community and its engineering consultants, contractors, and vendors.

For example, MRF projects have recently incorporated certain aspects of full service procurement in the D–B–B approach. Instead of bidding on individual pieces of equipment, the entire equipment process line is bid as a package. This minimizes the number of contractors which the community's construction manager must deal with

Solid Waste Recycling and Processing. DOI: http://dx.doi.org/10.1016/B978-1-4557-3192-3.00008-7

and provides a mechanism for sharing the risk of project performance with the contractor. Design of the remaining "wraparound" facilities is the responsibility of the engineer and is secured under a general contractor bid.

Design, Build (D–B) Approach

In a D–B approach, a single entity is awarded a contract for design, construction, equipment, and start-up of the project. The turnkey contractor selects the equipment and supplies to be used and may either design and construct the project itself or subcontract portions of the work. In either case, the contractor assumes sole responsibility for construction of the project. Upon completion of construction, start-up, and successful testing, the recycling facility is accepted by the community which then assumes responsibility for the operation or contracts for the operation with a private firm. A modified turnkey approach may include a short-term (e.g., 1–5 years) agreement with the contractor to operate the project after acceptance.

Design, Build, Operate (D–B–O) Approach

An extension of the D–B approach is to assign total responsibility for project design, construction, equipment, acceptance testing, operation, and possibly ownership to a single entity or full service contractor. Under this approach, the community is provided with long-term operation and maintenance services. This procedure usually includes a contract for design, construction, and acceptance testing and another long-term contract for project operation and maintenance.

Table 8.1 summarizes the project procurement and operation responsibilities associated with each of the procurement approaches.

Procurement Procedures

The procurement procedure should provide a fair and equitable method of selecting a contractor who promotes public confidence and at the same time takes advantage of the competition which exists in the materials recovery industry in order to get the most beneficial arrangement for the community. Competition between contractors should not be limited to construction price alone but should also take into consideration the quality and reliability of the system proposed, performance guarantees, and long-term operational and maintenance expenses as appropriate. The procurement should take these elements into consideration when the community evaluates various proposals prior to selection of a contractor.

Competitive Negotiations

The use of the competitive negotiation method of procurement is appropriate for those situations where the item or service desired requires extensive discussions with offers to determine the fairness and reasonableness of offers and to establish the allocation of risks upon which priced proposals will be made. In the public works area, "competitive negotiation" is better known as the "request for proposals (RFP) process."

Table 8.1 Typical MRF Procurement and Operation Responsibilities

Project Responsibilities	Procurement Approaches		
	A/E	**Turnkey**	**Full Service**
Planning	C or E	C or E	C or E
Plant design	E	V	V
Preparation and issuance of plant specification	E	V	V
Construction supervision	E	V	V
Construction and equipment installation	V	V	V
Start-up	C or V	C or V	V
Operation	C or V	C or V	V
Ownership	C	C	C or V

Source: From Ref. [1]
C, community; V, vendor or contractor; E, Engineer as agent for the community.

Throughout the world, procedures available for procurement of public facilities or services are generally subject to statutory bidding requirements or local ordinances. There are four generally acceptable bidding procedures: (i) competitive sealed bidding, (ii) competitive simultaneous negotiations prior to issuance of the final RFP, (iii) two-step competitive negotiations after issuance of the final RFP, and (iv) sole-source negotiations.

Competitive Sealed Bidding

Competitive sealed bidding or tendering is the most common method used for acquiring supplies, services, and construction for public works projects. This method requires the preparation and issuance of an Invitation for Bids (IFB) containing detailed specifications and a purchase description of the desired item, service, or construction project. Upon receipt of bids or tenders, the community determines if the item or service being offered satisfies and is responsive to the requirements of the IFB and if the bidder is responsible. No change in bids is permitted once they have been opened. Once bid evaluation is completed, the award is made solely in an objective basis to the lowest responsive and responsible bidder.

Two-Step Competitive Negotiation or Simultaneous Negotiations

Under this procedure, an RFP, as opposed to an IFB, is issued. The RFP contains minimum technical and financial qualifying criteria, general system and performance specifications, the evaluation criteria to be used (price being only one), and the relative importance of each evaluation component, as well as the basic features of the risk allocation package including, in many instances, a draft contract. Financial and minimum technical qualifications of bidders can also be performed in a separate Request for Qualifications (RFQ) issued prior to the RFP. Based on the RFQ responses received, a short list of qualifying vendors is prepared. After the receipt of the draft RFP and an appropriate period to receive initial vendor comments and requests for clarifications, discussions are held with all qualified vendors to arrive at a final RFP which is issued and upon which the vendors are required to submit bids without modifications. This method allows for "apples-to-apples" comparison or evaluation of submitted bids.

The selected vendors are ranked based upon the comparative evaluation of their respective proposals, and the community then enters into final negotiations with the top-ranked proposer, usually for a specified time period, to arrive at the most desirable transaction for the community. If negotiations are not satisfactorily concluded within that time period, then the community is able to terminate negotiations with the first vendor and commence negotiations with the second vendor. Furthermore, in the unlikely event that two or more of the top-ranked vendors' proposals are virtually the same, the community would reserve the right to enter into competitive simultaneous negotiations with those vendors. By limiting the negotiating period with the selected vendors, the competitive negotiation process enables the community to structure a competitive procurement on a cost-effective basis in circumstances where the number of qualified vendors would dictate against use of the competitive simultaneous negotiation approach.

This process differs from competitive sealed bidding in two major respects. First, judgmental factors are used to determine not only the compliance with the requirements of the RFP but also their impact on the evaluation of competitive proposals. The effect of this is that the quality of competing proposals may be compared and trade-offs made between the price and quality of offers. Second, since comments and contract discussions occur after the issuance of the draft RFP, and changes in proposals may be made to arrive at a final RFP which is the basis for each vendor's offer, the final RFP reflects an overall transaction, which is most responsive to the community's needs and provides a competitive environment for procurement of vendor proposals. Final award is made to the offerer whose proposal is most advantageous to the community based upon price and specified evaluation factors set forth in the RFP.

The two-step competitive negotiation procedure is used chiefly in situations where the complexity of the system or service desired requires the preparation of detailed specifications by the community. This procedure incorporates features of both the competitive sealed bidding and competitive negotiation methods. The community may limit the number of offerers through an RFQ process. In step one, the draft RFP is issued and discussions are conducted separately with the vendors based upon their initial comments to ensure complete understanding by the vendors of what the community requires in order to enable offerers to be more fully responsive to the final RFP. This step is similar to the competitive negotiations, up to the issuance of the final RFP upon which unpriced proposals are submitted. Proposals are then evaluated based upon criteria set forth in the RFP, and the two or three vendors submitting the most responsive proposals are selected for the negotiations phase of this procurement process which is completed simultaneously.

Sole-Source Negotiation

Sole-source negotiation involves no competition and is used when the community determines that there is only one source for the desired supply, service, or construction project.

Table 8.2 Applicability of Procurement Approaches and Procedures

Procedures	Approaches		
	A/E	Turnkey	Full Service
Competitive sealed bidding	Appropriate	Not appropriate	Not appropriate
Competitive negotiation/RFP process	Possible	Appropriate	Appropriate
Two-step competitive negotiation	Not appropriate	Appropriate	Appropriate
Sole-source negotiation	Possible	Appropriate	Appropriate

Source: From Ref. [1]

Not all procurement procedures are applicable for use with the available procurement approaches. The competitive sealed bidding procedure is only appropriate for the D–B–B procurement approach since only in the D–B–B approach is the project broken down into well-defined, discrete components of equipment, fabrication, on-site construction, and operation. Similarly, the essence of successful turnkey or full service procurement involves, in addition to considerations of price, a careful analysis of vendor experience, technology, guarantees, financial well being, and the overall risk allocation reflected in the transaction. Such factors are subjective to some degree, and the community must be permitted to make qualitative judgments as are provided in the competitive simultaneous negotiation, two step competitive negotiation, and sole-source negotiation procedures. Table 8.2 summarizes the relationship between procurement approaches and procedures.

Procurement and Risk Posture

The basic difference between the available procurement approaches is the level of control the community can retain in the design, construction, and operation of a recycling facility and the subsequent allocation of risk. A section on risk assessment is provided later, and the following paragraphs provide a brief summary of key risk issues.

Full service procurement allows the greatest sharing of risk with the private sector. In a full service procurement, contractor guarantees are available for a maximum construction price, completion with a specified construction period, demonstration and maintenance of specified performance levels, and a fixed annual operating and maintenance expense, usually subject to a negotiated escalation formula. A D–B procurement contractor will typically guarantee a maximum construction price, completion within a specified construction period, demonstration of specified performance levels prior to the community's acceptance, and a limited warranty (generally 1 year) on workmanship and materials. In the D–B–B procurement approach, the risks which can be shared include completion within a specified construction period and demonstration of specified performance levels.

In each procurement approach, the guarantees are subject to occurrences beyond the contractor's control such as changes in law and force majeure events.

Depending on the final negotiated agreement, there may be a maximum limit on the contractor's aggregate contractual liability.

There is a direct correlation between the amount of risk shared with the private sector vendor and the overall price of the project. Since a full service contractor typically takes the greatest amount of risk, a full service proposal will be higher priced than the D–B–B procurement approach. Because the guarantees associated with the risk posture accepted by the vendor are conditioned on certain contract terms, it is essential that the procurement procedure also clearly defines these conditions and the community's risk position.

Analysis of Procurement Options

Determination of the procurement approach and procedure is one of the first decisions the community must make in the implementation. In making the procurement decision, the community must first establish what its goals are in implementing the project control, ownership, and risk allocation structure.

The community's position on project control is influenced by (i) traditional community practices in procurement of public work projects, (ii) the public's perception of community involvement in an environmentally sensitive issue, and (iii) the technical, managerial, and financial capabilities of the community. The community should evaluate its position in response to several questions:

- Does the community want to own the project?
- Does the community operate the project?
- Does the community have the technical and managerial resources needed to operate the project in an efficient manner?
- What level of involvement does the community desire in project design, technology selection, site layout, and architectural standards?

The D–B–B approach is particularly applicable if the community desires to own and operate the project. The approach allows the greatest degree of community control regarding design features and therefore offers the greatest response to public concerns which may arise. However, the D–B–B approach also requires the greatest commitment of community resources in development and operation of the project and exposes the community to potentially greater economic risks. The key construction risk concerns are cost overruns, lack of sole-source responsibility, and potential technical failure. These risks are manageable and can be mitigated by selecting a competent and experienced D–B–B firm and by selecting only proven technologies and equipment. In addition, a good construction manager can minimize scheduling and coordination problems.

Operational risks pose a more significant problem in the A/E approach. If proven and reliable technology and equipment is used, operational performance risks are minimized. However, with public operation of the project, a highly qualified technical staff must be maintained in order to assure that adequate maintenance and proper operation is performed. For communities that have similar public work activities, this does not pose a significant problem, and the project staff can be formed relying in part on previous experiences. An additional concern is the continuity of financial commitment to maintain an adequate staff, particularly in light of fluctuations in the community's economy and budgetary constraints, as well as potential changes in political leadership's philosophy or support.

The turnkey procurement approach would be applicable if the community desires to establish sole-source responsibility for project construction, yet wishes to keep the option of community operation of the project. The community relinquishes control of project construction once the plant specifications are determined, except for the cost of work change orders. However, because the community is responsible for providing long-term operation either through community staff or through contracted operators, it is in the community's best interest to monitor the design and construction. This additional monitoring expense, coupled with the profit margin for accepting sole-source construction responsibility, usually makes the D–B approach costlier than the D–B–B approach.

The key risk area in the turnkey approach is the long-term performance of the recycling facility. This risk can be mitigated in part by selecting only sound and proven technologies and financially secure contractors. However, because the contractor is not responsible for operations, his/her profit lies in efficient construction, potentially resulting in a higher proposed capital cost, and there is some risk that construction shortcuts could result in operational problems over the long term or in additional expenses which are not apparent when the plant is tested and accepted.

To partially alleviate this concern, a modified turnkey approach has been proposed which includes a short operating term (i.e., 1–5 years) by the contractor. However, this is still only considered to be a short-term solution, and due to the lack of substantial short-term operating experience in the United States MRF industry, the effectiveness of such a modified approach has not been demonstrated. If the community desires to operate the project, the same concerns described in the D–B–B approach apply. If the community elects to contract for operation of the facility with a company other than the turnkey construction contractor, only limited operational risks would be accepted by such an operator due to the operator's lack of control over the design and construction. Furthermore, disputes arising over the cause of operating problems may be difficult to ascertain and resolve.

The full service procurement approach is applicable if the community views operation of the recycling program as obtaining a service, similar to the services the community currently receives for waste disposal. The degree of control available to the community depends, to a great extent, on the ownership status of the project. If the public maintains ownership, the community can require the operator, as the community's agent, to respond to the public requests on operating standards or procedures. However, if the community establishes more control on operating procedures than is considered to be standard in the industry, the full service operator will take fewer operating risks.

If the community elects to allow private ownership of the project, the public's control is limited to predetermined contractual rights such as annual inspections, maintenance of specific permit criteria, and other similar areas. Response to public concerns would depend on the working relationship established between the community and the private owner, and the private owner may not be legally required to respond to all problems. With private ownership, the community must also recognize that the ultimate responsibility for waste disposal rests with the community in the event of vendor bankruptcy, insolvency, or shutdown.

In terms of risk allocation, the full service procurement approach offers the greatest opportunity for the community to share construction and operational risks with the private sector. Because one company has the responsibility for all aspects of the project—design, construction, and operation—the vendor is willing to guarantee certain performance levels during operation, an annual operating and maintenance fee, usually subject to an escalation formula, and material floor prices. These operational performance and cost guarantees provide the community a more balanced risk allocation posture than is normally available for D–B–B and D–B projects. However, the full service contractor may also expect to be paid a higher operating fee or have a greater profit expectation for assuming such risks. There are certain risks beyond the contractor's control (e.g., changes in law, force majeure events) which will still be assumed by the community regardless of the ownership of the facility. The chief concern regarding the full service approach is the long-term financial stability of the contractor.

Recently, new corporate sponsors with access to proprietary resource technology have entered the recycling industry. To protect the public's interests, a vendor with both a sound and proven technology, and the financial strength to backstop construction, performance, and operating guarantees, must be selected. The community must also establish a contractual right to conduct regular inspections and reviews of the facility, and in the event the full service operator's corporate sponsorship varies, to maintain an independent financial guarantee of the parent company and its assignees as well as a separate contractual right to assure the continued availability of any proprietary technology.

A summary of the advantages and disadvantages of the procurement options is provided in Table 8.3.

Ownership Issues

Ownership of a solid waste and recycling project is another important community policy decision. Although it is generally considered to be a public responsibility to protect public health and provide for the safe disposal of solid waste, there are several reasons why private ownership of a recycling project is attractive to various communities:

1. Communities which have relinquished the responsibility for solid waste collection and disposal to the private sector are also inclined to relinquish ownership and control of a recycling project.
2. Private sector businesses (haulers) who control solid waste flow through collection and/or disposal operations can develop a project as part of their continuing business venture.
3. Legal restriction (e.g., special labor regulations, bidding laws, contractual restrictions) or lack of cooperation between multiple communities makes it too expensive or impossible to maintain public ownership.
4. Federal tax benefits are available to the private sector and may be applied to subsidize the cost of the project.

Table 8.3 Advantages and Disadvantages of Procurement Approaches

Approach	Advantages	Disadvantages
D–B–B	• Potential lower price • Allows community control and input on design	• No single-source responsibility • Coordination of multiple contractors • Higher community risk exposure • No long-term performance guarantees
D–B	• Sole-source responsibility for design, construction, and initial performance	• No long-term performance guarantees • Operating expense risk remains with the community • Community controls operation
D–B–O	• Sole-source responsibility for design, construction, and operation • Greater risk sharing potential • Guaranteed long-term performance	• Higher costs associated with long-term operating risk • Reduced community control

Source: From Ref. [1]

The choice of public or private ownership is not always clearly apparent, and community decision makers must analyze various project aspects before making a decision. The principal areas to evaluate in making the ownership decision are (i) economics, (ii) residual value, (iii) project control, and (iv) project risk and financing security. The following discussion focuses on the differences inherent in public and private ownership in these four areas.

Aspects of Public and Private Ownership Decision Making

Economics

Assuming that the cost of construction and operation of recycling project is the same regardless of ownership, the principal reason for the attractiveness of private ownership has been the availability to use federal tax benefits to subsidize project costs. The use of equity contributions will reduce the annual debt service requirement, either through an initial contribution and smaller bond size or through annual contributions during the initial years of project operation. A portion of the private owner's return on this investment is from the available tax benefits. This is taken into consideration when the tipping fee for the project is set by the private owner. Computer models are available to estimate the vendor's return and to optimize the equity contribution. There is no assurance, however, that the community would be given full credit for the available tax benefits, and the worth of tax benefits to individual owner is difficult to determine during negotiations. In addition, an annual operating profit would be included as part of the operating and maintenance expense. Such an operating profit would be common to both public and private ownership if a private company operates the project.

Residual Value

The term of a service agreement or loan agreement is typically 10–20 years; however, the useful life of a project (with proper repair and maintenance) may be in excess of this term. If properly maintained with replacement of parts on a regular schedule, the residual value of the project may be the same as the original construction price. When making the ownership decision, the community should take the following into consideration:

1. Barring any major technological breakthrough, the project would still offer an effective materials recovery system.
2. The community would, in all probability, continue to rely heavily on operation of the project as the primary means of commercial waste recovery.
3. For continued use of a privately owned project after the initial term of the service agreement, the community would be required to either renegotiate the service agreement or purchase the project.

Control

Solid waste projects are particularly vulnerable to public scrutiny regarding "environmental concerns," and some public officials prefer to have more extensive control over the operation of a project than is afforded by private ownership in order to satisfy these public concerns. With public ownership, the community has control over all aspects of project construction and operation. Some public officials prefer to distance themselves from public involvement in such projects and prefer private ownership. However, as previously discussed, the community could only have limited control over the operation of a privately owned project. This typically includes only the rights to inspect the plant and requires periodic tests to demonstrate guaranteed performance levels.

It becomes a subjective decision for the community to weigh the public's reaction to project control when making the ownership decision.

Risk and Financing Security

The relationship between procurement options and risk was discussed above. With regards to ownership, the risk posture under a full service procurement approach is basically the same with either public ownership or private ownership. In private ownership, however, the owner may be willing to assume a portion of the risk of certain unforeseen situations. The owner's risk position is generally limited to providing additional capital in the same proportion as its initial equity contribution and is generally limited in amount to the available tax benefits. In addition, in a private ownership arrangement, the community may be asked to take the risk of loss of anticipated tax benefits due to changes in tax laws (i.e., tax indemnification).

In terms of financing security, public ownership backed by a general obligation pledge would provide the most desirable investor security arrangement. If revenue bonds are used under either public or private ownership, the vendor may be willing to supply a corporate guarantee or letter of credit (LOC) support as additional security enhancement for the bonds. The credit worthiness of the corporation and the extent to which limited subsidiaries are used are factors in the investor's security analysis.

Risk Assessment

Risk is the possibility that an event that has detrimental impact on the project will occur. A risk assessment is an evaluation of possible risk events, the impact of a risk occurrence, and methods by which risks can be reduced/mitigated or shared/allocated. The result of a risk assessment is a definition of the community's risk and a determination of the implementation strategies which best reflect the community's risk posture; the community risk posture is defined in the contractual arrangements, procurement approach, and financing structure selected for a project. The following section provides the community with an overview of project risks and proposes a risk posture for consideration by the community. The community's actual risk posture should be determined once a "go/no go" decision on project implementation is made.

Risk Events, Impacts, and Allocation

Risk events can be categorized into the following general areas: (i) waste stream, (ii) project construction, (ii) project operation, (iv) materials market, and (v) financial and contractual matters. Risk events can be further categorized as to the cause of the event (i.e., community-caused, contractor-caused, or uncontrollable events) and appropriate mitigation or risk sharing measures taken. It is difficult to assess the probability of the occurrence of a risk event. However, events which are caused by the action or inaction of the community or caused by circumstances beyond the contractor's reasonable control will be the responsibility of the community. Events caused by or in the direct control of the contractor will be the responsibility of the contractor. Table 8.4 outlines typical risk sharing under the various procurement options.

The results or impacts of a risk occurrence are primarily financial in nature and can include one or a combination of the following situations:

1. Cost overruns and increased capital requirements.
2. Inability to use the project for commercial recycling and subsequent decrease in landfill capacity due to continued landfilling or increased nonprocessed waste or residue with increased landfilling requirements.
3. Increased operating and/or maintenance costs.
4. Lost or lower than anticipated materials revenues.
5. Lost or lower than anticipated tipping fees.
6. Temporary or permanent MRF shutdown.
7. Failure to make debt service payments.

The probability that a risk event will occur can be reduced or mitigated by the following:

1. Selecting only proven and reliable technologies.
2. Selecting creditworthy contractors with demonstrated performance and management skills.
3. Proper planning for project implementation, such as selecting project design and operating criteria to compensate for variations in waste composition and supply, providing screening measures for hazardous waste, providing protection for explosions, floods, and other force majeure events, and making provisions in the design for additional air emission equipment.
4. Legislative control of waste deliveries.

Table 8.4 Alternate Procurement Approaches Risk Sharing[a]

Risk element	D–B–B	D–B	D–B–O	
			Public Ownership	*Private Ownership*
Waste Stream				
Waste Availability				
Failure to deliver	Community[b]	Community	Community	Community
Haulers strike	Community	Community	Community	Community
Competitive diversion	Community	Community	Community	Either
Flow control failure	Community	Community	Community	Community
Waste Composition				
Evolutionary change	Community	Community	Either[b]	Either[b]
Change in law	Community	Community	Community	Community
Project Construction				
Contractor Fault				
Design errors	Community	Vendor	Vendor	Vendor
Construction strike	Community	Community	Community	Either
Subcontractor failure	Shared[b]	Vendor	Vendor	Vendor
Cost underestimated	Shared[b]	Vendor	Vendor	Vendor
Equipment performance failure	Vendor	Vendor	Vendor	Vendor
Uncontrollable Circumstances				
Subsurface conditions	Community	Community	Community	Community
Force majeure	Community	Community	Community	Community
Change in law	Community	Community	Community	Community/ shared
Inflation increases	Either	Either	Either	Either[c]
Tax law change	–	–	–	Shared[b]
Project Operation				
Contractor Fault				
Excessive downtime	Community	Vendor	Vendor	Vendor
Damage by waste	Community	Shared	Shared	Shared
Equipment underperformance	Community	Vendor	Vendor	Vendor
Uncontrollable Circumstances				
High O&M cost	Community	Either	Either	Either[b]
High inflation	Community	Community	Community	Either[b]
Change in law/regulation	Community	Community	Community	Community
Lower material revenues	Community	Shared	Shared	Either

Source: From Ref. [1]

[a]Community fault events are always considered community responsibility. "Either" reflects community or vendor responsibility; "Shared" reflects partial community and vendor responsibility.

[b]These fault events are usually resolved by negotiations.

[c]The risks of construction cost increasing during the operation period may be handled through a fixed price contract (vendor risk) or escalating payment (community risk).

5. Establishing reserve funds for major repairs.
6. Securing reasonable insurance policies.

The degree to which risks can be shared by the community with a private contractor depends on the type of procurement approach selected. Project risks will be accepted by private contractors; however, adequate compensation for such risk sharing will be expected.

The community's risk posture should take into account that in order for any risk sharing arrangement to work:

1. The risk allocations must be equitable to all parties.
2. A good working relationship between the community and the vendor must be established.
3. Natural positive incentives (e.g., revenue sharing) should be used to ensure performance.
4. Flexibility should be provided in long-term operating agreements so that a contractor can maintain a minimum profit during hardships, thus avoiding bankruptcy or insolvency.

Waste Flow Considerations

One of the more critical issues facing public officials pursuing solid waste and recycling projects is what is commonly termed "waste flow control." In essence, each community must be able to assure those who will be operating its facility and the financial underwriters for such a project that the solid waste or recyclables generated from residential, commercial, and industrial establishments within the community will be available on a long-term basis to supply the facility. Without control of the solid waste stream, there is the potential for solid waste from the community to be diverted to other processing or disposal facilities in the region. This would be an unacceptable situation because revenues from both tipping fees and the sale of electricity, steam, and/or recovered materials are the main collateral for financing the construction and long-term operation of such facilities.

Waste flow control has been a controversial issue in recent years in the United States. Simply put, the waste-to-energy (WTE) industry and local governments favor waste flow control but solid waste haulers and the waste recycling industry oppose. The latter group argues against the imposition of monopolistic waste flow control measures by local government for WTE facilities because these measures threaten to reduce their traditional access to recyclable materials available for extraction from the waste stream, thus leading to a reduction in revenue and profit. Representatives of waste haulers have asserted that diversion of materials from a community's waste stream via a recycling program typically benefit rather than impair the financial integrity of WTE programs because the size and capital costs of such facilities could be reduced through initiation of waste flow reduction programs.

The hauler's argument has been rejected by many communities and the investment community. Spokesmen for these groups have argued that the financing of WTE facilities cannot take place without the long-term assurance on the part of government that a community's solid waste is committed for delivery to the WTE facility. Without a guaranteed delivery of waste, the investment community asserts that the interest rate paid by WTE plant owners for project financing would increase dramatically. Furthermore, some representatives of local government have asserted their rights to prohibit the informal scavenging of materials at the curbside because of public health and safety considerations. Some communities in recent years have attempted to take a middle course by enacting waste flow ordinances with commitment for WTE facilities, while at the same time encouraging the development of a strong recycling industry in their community. These models appear to be working well. The bottom line is waste reduction and the development of WTE projects need not be incompatible.

Flow Control Mechanisms

Addressed below are the three basic types of waste flow control mechanisms prevalent in the United States: (i) legislated regulation supplemented by enforcement, (ii) contractual, and (iii) economic or cost incentives.

Waste Flow Control Through Legislation/Regulation

Local government agencies in the United States are empowered to exercise legal or regulatory authority over the collection, removal, and disposal of solid waste generated by citizens and businesses in their areas of jurisdiction. Courts have long upheld the rights of governments to adopt reasonable regulations in this regard since property rights are considered to be superseded by local government's police powers. Most of court cases involving the management of municipal solid waste were decided by jurists at the turn of the twentieth century on the premise that regulatory authority was essential to maintain public health and safety because without such control, solid waste would become a nuisance and/or present a health hazard to neighboring property owners.

As solid waste facilities were being implemented in the United States in the early 1980s, there was some concern that solid waste flow control would be deemed unconstitutional to preserve the free flow of interstate commerce. Perhaps, the most important legal decision regarding solid waste flow control by local government during that era involved the City of Akron, Ohio. Under the terms of the convenants for the $46,000,000 bond issue to fund its 1000 ton-per-day WTE facility, the City of Akron was required by its bond underwriters to enact an ordinance that would do the following: guarantee the supply of solid waste to the facility by prohibiting the establishment of alternative solid waste disposal facilities, require all garbage collectors within both the City of Akron and Summit County, Ohio, to deliver all waste acceptable for processing at the plant (including recyclables), and require all collectors to pay a tipping fee when they deposited solid waste at the facility. Violation of this city ordinance by a solid waste collector would result in possible loss of his/her license to collect wastes and subject him/her to criminal penalties.

Prior to enactment of this ordinance in the late 1970s, private collectors in the City of Akron and Summit County were able to shop around for solid waste disposal sites with the best disposal price and also were able to recover and sell valuable recyclables from the solid waste stream they collected before taking the remainder to the landfill. The imposition of this waste flow control ordinance in metropolitan Akron markedly curtailed the operations of the private collectors and landfill owners by substantially reducing their incomes.

In the lawsuit *Glenwillow Landfill, Inc. v. City of Akron, Ohio*, 485F. Supp. 671 (ND Ohio 1979), these groups argued in Federal District Court that the City solid waste control ordinance violated due process; took private property in violation of the Fifth Amendment of the US Constitution; illegally restrained interstate commerce allowed under the Commerce Clause of the US Constitution; and violated the Sherman Anti-Trust Act. In ruling for the City of Akron, the court found that the ordinance was a proper exercise of the city's police powers, and as such did not result in a "taking" for which compensation must be paid under the Fifth Amendment of the US Constitution.

The District Court's ruling was appealed to the US Court of Appeals, Sixth Circuit. In *Hybrid Equipment Corp. v. City of Akron*, Ohio, 654F. 2d 1187 (6th Circuit 1981), the Circuit Court upheld the City of Akron's waste control ordinance. Citing two US Supreme Court cases: *California Reduction Company v. Sanitary Reduction Works*, 199 U.S. 306, 26S.Ct. 100, 50L.Ed 204 (1905) and *Gardner v. Michigan*, 199 U.S. 325, 26S.Ct. 106, 50L.Ed 212 (1905), the Court found that the City had not violated the due process and taking clauses of the US Constitution, and the ordinance was a proper exercise of the traditional exercise of local governments police powers. The Court also ruled that the City's actions did not seriously burden interstate commerce, and that the City was exempt from the Sherman Anti-Trust Act. The plaintiffs in this case then appealed to the US Supreme Court.

However, before this case reached the Supreme Court, the Court had handed down a ruling in *Community Communications Company, Inc. vs. City of Boulder, Colorado*, 455, U.S. 40, 102S.Ct. 835, 70L. Ed 2d 810 (1982), holding that a municipality can be held responsible for violations of the federal antitrust laws unless it is acting pursuant to "...a clear and affirmatively expressed state policy" permitting such restraint of trade. In light of its decision in the Boulder case, the Supreme Court overturned the judgment against the plaintiffs in the Akron case (solid waste haulers) and sent the case back to the Circuit Court to reconsider the issue of state action exemption for local government under the Sherman Anti-trust Act. The Sixth Circuit remanded this case to the Federal District Court for disposition.

In reviewing the facts of the case, the District Court ruled that the City of Akron was exempt from antitrust liability because the City was acting in furtherance of "clearly articulated and affirmatively expressed" policies of the State of Ohio in the financing of waste disposal facilities. The Court found that the State Legislature of Ohio had contemplated the use of anticompetitive measures to ensure the financial viability of its waste disposal facilities. Consequently, the Court reasoned that as long as local government was acting pursuant to a clearly articulated and affirmatively expressed state policy that indicates an intent of the legislature to displace competition with regulation, local government would be exempt from antitrust liability. Furthermore, the Court, relying on *Town of Hallie vs. City of Eau Claire* (No. 82–1715, Slip Op, 7th Circuit Court February 17, 1983) held that the active state supervision requirement for antitrust immunity does not apply to municipalities engaged in a traditional municipal function authorized by the state.

Prior to the final disposition of the Akron case, the financial and legal advisors to communities who were hoping to implement WTE facilities insisted that some sort of legal mechanism be used to confer state immunity from antitrust actions upon local government. To achieve such immunity, however, it was believed that local government needed to meet the "active" state supervision test. Some local governments have attempted to demonstrate such state oversight of their WTE programs by having the state, through special legislation, officially delegate the power of supervision to them. Others through special legislation have reaffirmed the existing state supervisory powers over its WTE systems, including the issuance of permits and periodic reviews.

The Akron decision makes it clear that waste flow control ordinances in support of WTE facilities must be drafted by a knowledgeable team of lawyers and other professionals. The ordinances must balance the needs of government to assure secure waste supplies for its WTE facility against the legitimate economic concerns of waste haulers and the recycling industry who seek to remove recyclable materials from the waste stream. Many local governments, while enacting strong flow control ordinances, now permit the recovery of recyclable materials from the waste stream prior to delivery to the WTE plant. Nonetheless, solid waste flow control continued to be litigated in the United States for another decade.

In 1994, the US Supreme Court issued a far-reaching and landmark legal opinion regarding solid waste flow control. In *C&A Carbone, Inc., et al., v. Town of Clarkstown, New York* (1994), the Court deemed a local flow control ordinance in New York unconstitutional because it violated the Commerce Clause of the US Constitution by driving both local and out-of-state competitors away from the local market. The Town of Clarkstown had retained a private contractor to build a waste transfer facility and enacted an ordinance requiring that all solid waste generated within the Town be directed to the transfer facility. (Tipping fees at the transfer station were set higher than disposal costs then-prevailing in the private market.) The Town had financed the transfer station and planned to be paid back from tipping fees generated there. The Supreme Court struck down the ordinance on the basis that solid waste was a commodity in commerce and that the Commerce Clause invalidates local or other laws that discriminate against such commerce on the basis of its origin or its destination out-of-state.

The Carbone decision was soon applied to solid waste management programs in New Jersey in 1997, in the case of *Atlantic Coast Demolition & Recycling, Inc. v. Board of Chosen Freeholders of Atlantic County, et al.* 112F.3d 652 (3d Cir. 1997 ("Atlantic Coast"). New Jersey is one of the most densely populated states in the nation. Coupled with the scarcity of available land for landfills, its large waste-generating population, and the State's geographic location relative to other large metropolitan centers (Philadelphia and New York), New Jersey mandated costly upgrades to waste management systems in the state over the last three decades to ensure adequate, safe, and effective disposal capacity. As such, the state plan called for construction of 13 new solid waste WTE facilities, while at the same time, the State and local communities had to grapple with the closure of more than 500 Landfills with poor environmental records that failed to meet increasingly stringent regulations, which were being imposed nationwide by the US Environmental Protection Agency (EPA). At the same time, the State Legislature was looking to correct the long history of anticompetitive conduct by the state's solid waste industry through rate regulation and state certification of waste collectors.

In April 2007, the US Supreme Court made a landmark decision regarding solid waste flow control in *United Haulers Association v. Oneida–Herkimer Solid Waste Management Authority* 550 US (2007). United Haulers had sued the New York counties of Oneida and Herkimer claiming that county ordinances regulating the collection, processing, transfer, and disposal of solid waste violated the Commerce Clause. The flow control regulations enacted by the two counties required all solid wastes and recyclables generated within Oneida and Herkimer counties to be delivered to one of several waste processing facilities owned by the Oneida–Herkimer Solid Waste Management Authority. United Haulers had argued that these ordinances burden interstate commerce by requiring garbage delivery to an in-state facility, as this restriction necessarily prevents the use of facilities outside the Counties and diminishes the interstate trade in waste and waste disposal services. United Haulers had submitted evidence that the flow control ordinances increased the cost of waste transport disposal from between $37–55 per ton without flow control to $86 per ton with flow control.

By a 6–3 decision, the Court ruled in favor of the two counties, upholding solid waste flow ordinances that required waste haulers to deliver their trash to a publicly operated processing site. The justices disagreed with United Haulers, stating that the counties' flow control ordinances, "which treat in-state private business interests exactly the same as out-of-state ones, do not discriminate against interstate commerce." In stark contrast to the previous *Carbone* decision, a majority of the justices argued that this case was different because it dealt with a publicly owned solid waste facility, which benefits the local "government's important responsibilities include protecting the health, safety, and welfare of its citizens." The precise scope and impact of the Supreme Court's decision is currently unclear. Since the decision was announced, a few counties and municipalities have begun adopting solid waste flow ordinances.

Contractual Control of Waste Stream

Rather than resorting to the enactment of waste stream control legislation, local government can assure adequate quantities of solid waste for its WTE facility through contractual arrangements. This is accomplished when local government enters into one or more long-term contracts with other local governments and/or private collectors to deliver solid waste to a designated WTE facility. This method of voluntary contractual commitments can be particularly effective to secure an adequate load quantity of solid waste for the facility.

Contracting has been successfully applied to several WTE facilities in recent years to assure long-term waste supplies. For example, refuse for the Northeast Massachusetts Resource Recovery Project, located in North Andover, Massachusetts, is delivered to the facility by 22 municipalities that have signed 20-year "put-or pay" agreements for waste disposal and eight commercial haulers that have signed private hauler agreements. These "put-or pay" agreements require each community to deliver a guaranteed annual tonnage which can be adjusted yearly within certain limitations. These communities are assessed penalties for shortfalls or excesses below or above their contractual guarantees. Private haulers are also assessed penalties if their deliveries are below or above their contractual guarantee. In this way, the WTE facility can assure that adequate quantities of solid waste are delivered to meet its financial commitments to the local communities, the facility operator, and the bondholders.

Economic Incentives for Waste Stream Control

Waste stream control can also be achieved by local government through economic incentives. For example, the operator of a recycling facility can attract solid waste from both public and private collectors by charging a lower tipping fee than alternate disposal methods such as sanitary landfills. (In some cases, for instance where the sale price of electricity is relatively high, the facility owner may elect to charge no tipping fee.) In this case, private haulers would be attracted to the facility since they would have little or no economic incentive to dispose of their solid waste at sanitary landfills or other facilities located further away.

In order to achieve this sort of economic control over solid waste for its facility, the community must be willing to subsidize the loss of project revenues that might otherwise be realized through tipping fees with funds from other source(s), such as the general fund, a user fee, or a tax. For example, a user fee for solid waste disposal can be established for different residential, commercial, and industrial accounts, whereby the proceeds from this fee could be used to offset the artificially low tipping fee at the WTE facility. Some communities have also used revenue from property or other local government taxes to subsidize the tipping fee at their WTE facilities. Use of property taxes for this purpose, however, may be viewed by the investment community as a general obligation of the community and could result in a reduction of its bond rating.

Summarized below is a brief review of recent court cases across the nation involving solid waste flow control issues.

Carbone Revisited

In 2008, Rockland County, New York, enacted a solid waste flow control ordinance, whereby solid waste, yard waste, and residential recyclables, which are placed on the curb form pickup, must be delivered to the county's designated facility. This ordinance also requires that C&D debris be placed in a dumpster, and similarly, commercial solid waste and recyclables, must be delivered to this facility. The ordinance excludes recyclables generated by big-box retailers, as well as those which initiated operations of a recycling program prior to 2008.

C&A Carbone, Inc. and other waste firms doing business in the County filed suit in Federal District Court, alleging that the flow control ordinance violated the plaintiff's rights to deliver and dispose of locally generated waste streams. This complaint alleged that the County violated the interstate commerce clause of the US Constitution, took property without compensation, unlawfully impaired contracts, and violated the New York State Environmental Quality Review Act. The plaintiffs filed a motion seeking $75 million in damages plus attorney's fees. In its complaint, Carbone argued that the new ordinance is a replay of the previous Carbone case, whereby the US Supreme Court struck down the County's ordinance favoring a preferred private operator of the County's transfer station, Clarkstown Recycling Center.

In September 2011, the US District Court judge denied an initial motion (*C&A Carbone, Inc. v County of Rockland*, No. 08 Civ 6459 (KMK), S.D.N.Y) by the County to dismiss the complaint based on the allegation that the County, the Town of Clarkstown, and the recycling center conspired to favor a private facility at the expense of other private facilities. The case is proceeding with discovery with the final outcome of the case perhaps many years away.

City of Dallas, Texas

In September 2011, the City of Dallas enacted an ordinance requiring all commercial waste collected within the City be taken to the City's landfill. The intent of the ordinance was to assure the City a source of revenue to create waste management facilities and infrastructure and to support other environmental programs [2].

Previously, solid waste from commercial waste generators had been handled by private haulers who are bound by the terms of nonexclusive, long-term franchise agreements with the City. These agreements did not limit where solid waste and recyclables could be transported or disposed of. Roughly about half of the city's waste stream currently ends up in non City landfills or at recycling facilities within or outside the City.

The haulers filed a complaint against the City alleging that the ordinance was an unlawful impairment of the franchisee's rights to collect and dispose of waste and recyclables at a facility of their own choosing, violated the State law, the City's Charter, and Federal law. The plaintiffs sought a temporary and permanent injunction to prevent the ordinance from taking effect.

On January 31, 2012, the US District Court issued an order enjoining the City from implementing the flow control ordinance (*National Solid Waste Management Association et al. v. City of Dallas*, No. 3:11-CV-3200-O, N.D. Tex.). The judge found that the ordinance impaired the franchisee's contract rights and had a significant financial impact upon them. Further, the judge found that the only reason why the City had enacted the ordinance was to "increase the balance of the City's General Fund."

City of El Paso, Texas

The City closed their landfills in 2004. In order to meet waste disposal needs, the City signed an agreement with Waste Connections, Inc. (WCI), a national waste hauling and disposal contractor. In return, the City received access to WCI's landfill at reduced tipping fees while WCI was granted the right to provide commercial waste collection and disposal services within the City for 10 years.

In 2010, the City passed an ordinance requiring commercial solid waste generators to contract with franchise collectors. This agreement required that these haulers must bring waste to a designated solid waste facility for disposal. In response, WCI alleged that this new ordinance was contrary to their preexisting agreement with the City. However, the City Attorney issued an opinion indicating that certain portions of the WCI agreement were illegal under Texas law.

WCI filed suit in State court (*Waste Connections, Inc. v City of El Paso County Court*, No. 2010–4476) seeking a judgment blocking this new waste flow control ordinance and validating their 2004 agreement with the City. Subsequently, the City backed off on its flow control ordinance, putting waste flow control on hold until 2014 (when the City's agreement with WCI expires). In September 2011, WCI withdrew its lawsuit, but reserved its right to refile it in the future [2].

Horry County, North Carolina

A 2011 Horry County, North Carolina, case involves a series of legal allegations such as equal protection, due process, impairment of contracts, and restrain on interstate commerce. In 2009, Horry County enacted a local ordinance requiring locally generated solid waste to be disposed at a facility designated by the county's solid waste authority. Subsequently, a landfill operator Sandlands C&O, LLC and an affiliated hauler, Express Disposal Service, LLC, sued the County asking the court to invalidate the local County ordinance and recover $25 million in damages. The plaintiffs also alleged that the ordinance is preempted by State law, improperly delegates authority to the solid waste authority, interferes with current and prospective contract relations, takes property without just compensation, and violates the State's unfair trade practices act [2].

On September 19, 2011, the South Carolina Supreme Court ruled that the state solid waste law and related regulation do not preempt the County's ordinance (*Sandlands C&O et al. v. County of Horry, Supreme Court Opinion*, No. 27042). In a summary opinion, the Federal District Court rejected all of the plaintiff's arguments that the County's flow control law was invalid (*Sandlands C&O, LLC, et al. v Horry County, et al.*, No. 4:09-cv-01363-TLW, D.S.C.).

Oswego County, New York

In 2009, the County enacted a flow control law covering locally generated solid wastes. A local waste hauler, JWJ Industries, Inc. (JWJ), owns a solid waste transfer station, which enables the firm to accept C&D debris and solid waste from many sources. JWJ filed suit against the County alleging that the ordinance was an unconstitutional taking of property without just compensation, denial of due process, and equal protection. Importantly, the suit did not allege impacts regarding interstate commerce [2].

The Federal District Court initially issued a temporary restraining order against the County, but after subsequent hearing denied the plaintiff's request for a preliminary injunction (*JWJ Industries, Inc. v Oswego County*, No. 5:09-CV-0740 (NPM/DEP), N.D.N.Y). The judge later ruled that the County must correct vague and contradictory wording in the ordinance about which wastes are covered under the ordinance and which are not.

Subsequently, the ordinance has been amended to allow JWJ to take in-county waste to its facility to recover recyclables, provided that the residue goes to the County solid waste facility. JWJ can accept any waste originating outside the County.

References

[1] Rogoff MJ, Screve F. Waste-to energy technologies and project implementation. Waltham, MA: William Andrews; 2011.

[2] Solid Waste Association of North America. Litigation affecting management of municipal solid waste: Silver Spring, MD; 2012.

9 Case Studies

Over the past decade, hundreds of communities across the United States have implemented many different kinds of recycling programs. Some have concentrated on the residential sector by emphasizing curbside collection programs for single-family homeowners. Others have attempted to minimize their capital and operating expenditures for collection programs through the use of single-stream recycling. Still other communities have concentrated on diverting organic materials from their landfills to mulching or composting facilities as they move to zero waste.

As we have shown throughout this book, there is no magic formula for success in a recycling program. Each community must balance its recycling goals against available resources (time, money, and manpower), the cost and availability of alternative solid waste reduction and disposal systems, and institutional obstacles. We believe that the case studies discussed in the pages which follow illustrate creative programs whose decision makers had to balance these factors against the light of political reality. These represent just a sample of the many excellent programs which are currently in operation on a day-to-day basis.

Delta Air Lines In-Flight Recycling Program

Goals

Delta initiated its in-flight recycling program in 2007 on domestic flights arriving into Atlanta. The program has grown as the airline recycles aluminum cans, plastic cups and bottles, newspapers, and magazines.

Program

In-flight recycling operates in 28 US cities and one international location through a combined effort of flight attendants, catering staff, and cabin services. It was expanded to Punta Cana International Airport in Dominican Republic as the first international in-flight recycling station.

The education program includes computer-based training, a chapter in the in-flight handbook for policies and procedures, in-flight boarding screens, entertainment videos, Delta Net articles, flight attendant lounge visits, and employee incentive programs.

Habitat for Humanity homes are constructed with rebates generated from recyclable materials from the in-flight program through Delta's Force for Global Good.

Solid Waste Recycling and Processing. DOI: http://dx.doi.org/10.1016/B978-1-4557-3192-3.00009-9

Performance

1.1 million pounds of passenger materials were collected in 2012; 7 million pounds have been collected since the start of the program in 2007.

Additional Programs

Delta also operates an Employee Recycling Center at its Atlanta headquarters. This facility currently accepts aluminum and tin cans, plastic bottles and jugs, newspapers, magazines, phone books, junk mail, paperboard boxes, cardboard boxes, and glass. Rebates from these recyclables are donated to Delta's Employee and Retiree Care Fund.

Delta also has a carpet recycling program that has diverted 677,180 pounds of worn aircraft carpet through its partnership with Mohawk Aviation Carpet and Mohawk Recovery program since 2008. This equates to about 216,698 yd^2 or 45 acres.

Awards and Recognition

Member of Dow Jones Sustainability Index

2009 Green Save Award from Atlanta Business Chronicle for In-flight Recycling

2010 Green America Report on Airline Recycling Programs

Figure 9.1 Delta Airlines Recycling Program. (From Helen Howes)

Contact:

Helen Howes

helen.howes@delta.com

City of Lund, Sweden

Goals

Sweden's 290 municipalities are responsible for collecting and disposing of household waste, except for the product categories covered by producer responsibility. Municipalities are also responsible for collecting dry-cell batteries.

Sweden has producer responsibility for end-of-life packaging, cars, tyres, recycled paper, and electrical and electronic products. This responsibility imposes on anyone manufacturing or importing a product a duty to ensure that it is collected, processed, and recycled. The aim is to persuade producers to reduce waste quantities and ensure that waste is less hazardous and easier to recycle.

Producers have established "material companies," which contract service providers to arrange the actual waste management and ensure that targets are met. Collection and recycling are financed by the charge allocated by each material company to the products covered by producer responsibility.

Households have an obligation to separate and deposit waste at available collection points. They also have an obligation to comply with municipal waste management regulations.

To comply with these national requirements, the City of Lund has introduced the "4 fractions model" for household waste. This solution allows a collection of a pure high-quality fraction. The simplicity and close proximity of the system enables a higher quantity of materials, and the customers' motivation to recycle rises. Higher qualities and quantities means less work at the recycling companies, and most important, less waste that needs to be incinerated. The City's overall goal is 80% covered by doorstep collection (residential and multifamily).

Current Programs

The City currently has about 110,000 inhabitants—36% residential and 64% multifamily.

Residential areas are serviced with 2 × 370L bins with fortnightly collection, each bin on alternate weeks. The system uses two bins that are divided into four compartments, where every compartment has a designated waste material. This allows for collection of eight different fractions. Emptying is done with four compartment vehicles that empty all four fractions in the bin in one single cycle. The vehicle has divider walls, thus making mixing of fractions virtually impossible.

Multifamily recyclables are collected either in bins or in containers.

Green waste—households can contract a green waste fortnightly collection during March to December.

Performance

Collection:

- Municipal waste, 203 ton/year
- Recyclables, 123 ton/year

Treatment:

- 40% recycling (including biological treatment)
- 60% energy recovery

Contact:

Lena Wallin
Lunds Renhållningsverk
lena.wallin@lund.se

The Ohio State University (OSU) Zero Waste Program

Goals

In 2011, as part of its overall commitment to sustainability and reducing our carbon footprint, Ohio State set a goal to move Ohio Stadium toward zero waste. This is, diverting at least 90% of total game day waste generated from the region's landfill. The program includes sorting all recyclables and sending all potentially compostable organics to a compost facility. OSU has switched to compostable products (e.g., plastic coffee stirrers and sandwich toothpick holders to compostable wood products).

Participants

These efforts required significant and comprehensive partnerships across campus and the community. Ohio State partnered with the following organizations and groups for the zero waste program:

- Alpha Tau Zeta Chapter of FarmHouse International Fraternity (Compost sorting)
- D&D Cleaning (Compost sorting)
- The Ohio State University—Department of Athletics (Leadership and Operations)
- The Ohio State University—Energy Services and Sustainability (Leadership and Coordination)
- The Ohio State University—Facilities Operations and Development (Leadership, Support, Operations)
- IMG Sports Marketing (Sponsorship)
- Mid Ohio Foodbank (Food Donation)
- Ohio Department of Rehabilitation and Corrections (Recycling)
- The Ohio State University—President's and Provost's Council on Sustainability (Leadership and Funding)
- Price Farms Organics, LTD (Composting)
- PSI (Education and Outreach)
- The Ohio State University—NROTC (Operations)
- Rumpke (Operations—Hauling)
- Sodexo (Food services—Product conversion)
- Solid Waste Authority of Central Ohio (Support)
- Southeastern Ohio Correctional Institute (Recycling)
- Vera Institute of Justice (Support)

Costs in 2011

Change to compostable products—~$30,000
New containers and signage—$50,000
New team members at game day—$49,000
Marketing and communication—$5000

Performance

2011 reported 75.2% season diversion rate.
2012 reported 87.2% season diversion rate (season high 98.2%).

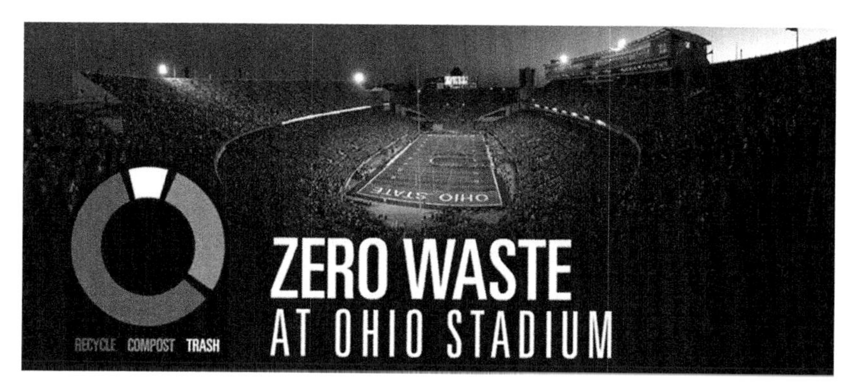

Figure 9.2 Ohio State University Recycling. (From Corey Hawkey)

Contact:

Corey Hawkey
Sustainability Coordinator
The Ohio State University
Hawkey.13@osu.edu

Toronto, Canada Recycling Program

Goals

Toronto is the largest city in Canada. The City's Solid Waste Management Services Division is responsible for municipal and some private sector waste disposal and recycling within the City. The City of Toronto has established a goal to divert 70% of its waste from landfills.

Programs

Recycling is mandatory for all City residents. Single-stream recycling is provided biweekly to single-family homes for the following items:

- Plastic bottles and jugs
- Milk/juice cartons and boxes
- Glass bottles and jars
- Plastic food jars, tubs, and lids
- Clamshell/hinged plastic containers
- Clear plastic food containers
- Metal cans/cardboard cans, aluminum trays
- Aerosol cans
- Metal paint cans and lids
- Foam food and protective packaging
- Plastic retail bags
- Bags, junk mail/gift wrap, newspapers, flyers, magazines, and so on
- Boxboard boxes
- Corrugated cardboard

Food[1] items include:

- Fruits, vegetable scraps
- Meat, shellfish, fish products
- Pasta, bread, cereal
- Dairy products
- Coffee grounds
- Soiled paper towels, tissues, soiled paper food packaging
- Paper plates
- Candies, cookies, cake
- Baking ingredients
- Household plants
- Diapers, sanitary products
- Animal waste, bedding
- Pet food

The City also provides weekly collection of multifamily recyclables (nine or more units). The same materials collected in residential program are also accepted from 340L wheeled plastic bins or 2–6 yd^3 front-end containers. Food waste collection is also offered.

Performance

- Recyclables: curbside participation rate—96%, 146,538 tons (2011)
- Food waste: curbside participation rate—89%, 100,663 tons (2011)
- Percentage of waste diverted: 49% (2011)

Single family: 64%
Multifamily: 20%

[1] Food waste (Green Bin Program) is collected separately from leaf and yard waste.

Figure 9.3 City of Toronto Curbside Program. (From Vincent Sferrazza)

Contact:
Vincent Sferrazza
vsferra@toronto.ca

Seattle–Tacoma International Airport (Sea–Tac)

Goals

Sea–Tac is the 17th busiest passenger terminal in the United States. The Waste Reduction and Recycling Program coordinates with airport staff, tenants, and business partners to reduce solid waste generation and landfill disposal. In 2009, Sea–Tac set a goal to recycle 50% of solid waste generated at the airport by 2014. We reached 30% in 2012.

Inside the airport terminal, passengers have access to recycling kiosks for recyclables and compostable materials at concourses, security check points, and in food courts.

Types of materials recycled:

- Beverage containers
- Mixed paper
- Cardboard
- Cooking oil
- Coffee grounds

- Batteries
- Printer cartridges
- Metals
- Wood
- Pallets
- Plastic films
- Food waste

Sea–Tac has installed 12 pairs of 30–40 yd^3 compactors with computer monitoring for "full alerts." Tenants have free use of the recycling compactor system, and they are charged for the use of the trash compactors.

Awards

2008–2012 Best Workplaces for Waste Prevention and Recycling, King County, WA.

Figure 9.4 Sea-Tac Recycling Program. (From Elizabeth Leavitt)

Contact:
Elizabeth Leavitt
Director of Planning and Environmental Programs
leavitt.e@portseattle.org

Seattle, Washington Recycling Program

Goals

The Seattle Public Utilities (SPU) operates the City's solid waste management system. City Council has set Seattle's goal to reach 60% recycling of municipal solid waste (MSW) by 2012 and 70% by 2025. Construction and demolition debris is excluded from this recycling goal.

Programs

Single-family residences receive a 64-gallon recycling cart for single-stream recycling, which includes the following:

- Newspaper, magazines, phone books, mixed paper
- Coated papers including hot drink cups
- Plastic tubs and bottles
- Plastic food containers and lids
- Plastic bags
- Metal cans
- Aluminum foil
- Scrap metal
- Glass bottles and jars
- Motor oil

Food and yard waste is required for all single-family households and collected on the same day as garbage. Acceptable items include:

- Meat, fish, and chicken
- Coffee grounds and filters, tea bags
- Fruits and vegetables
- Pasta, bread, grains, and rice
- Eggshells, nutshells
- Dairy products—milk, butter, cheese
- Food-soiled paper including pizza boxes
- Paper towels, napkins—kitchen only
- Paper plated—uncoated only
- Shredded paper
- Paper bags (uncoated) with food scraps
- Compostable bags
- Leaves, branches, twigs—up to 4 in. in diameter and 4 ft in length
- Plant and tree trimmings, grass
- House plans
- Small amount of sod—less than 60 pounds
- Holiday trees
- Multifamily complexes (four units or more) are provided recycling and food waste collection service with same list of materials as residential homes

Performance

Single family: 66.1 lb/month or 70.6% recycling rate (2011)
Multifamily: 29.6 lb/month or 28.7% recycling rate (2011)

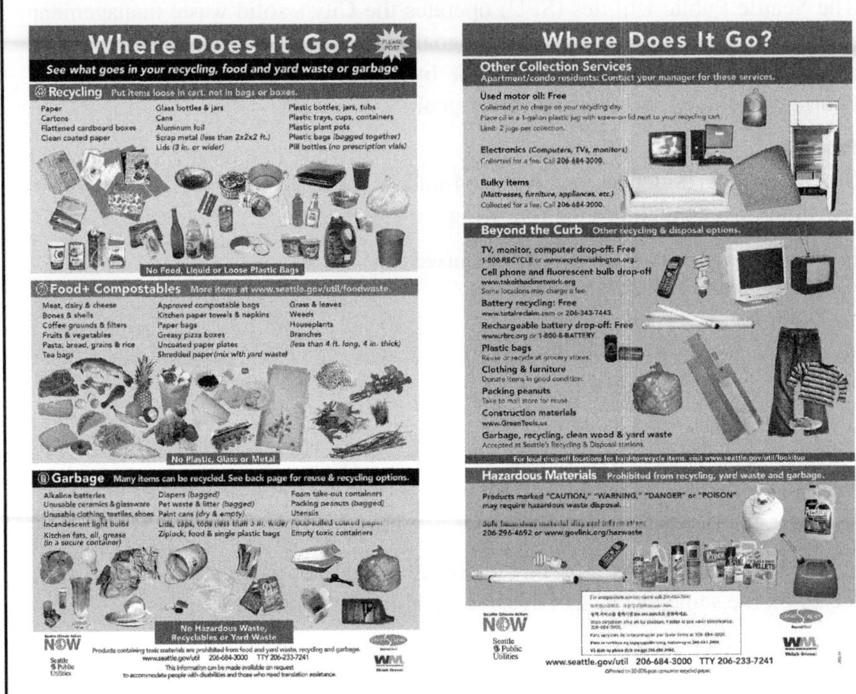

Figure 9.5 City of Seattle Customer Recycling Brochure. (From Brett Stav)

Contact:

Brett Stav
Sr. Planning & Development Specialist
Seattle Public Utilities
brett.stav@seattle.gov

City of San Francisco, California Recycling Program

Goals

The City's Department of the Environment (SF Environment) and Recology, the City's permitted hauler and processor run the City's solid waste programs. SF Environment has established a mandatory recycling and composting diversion program that requires everyone in the City to source-separate their recyclables, compostables, and trash into proper containers. The City's goal is to reach landfill diversion rate of 75% by 2010 and zero waste by 2020.

Programs

Recology provides a three-stream collection program using different colored bins:

- Blue—commingled recyclables
- Green—compostable organics
- Black—landfill waste

Single-family residents are provided 32-gallon bins (larger 64- and 96-gallon recycling and composting bins are available at no cost).

Acceptable recyclables include:

- Food scraps
- Plants
- Food-soiled paper
- Other (cotton balls, hair, fur, feathers, compostable food ware, wood crates, waxed cardboard, wooden chopsticks)

Multifamily complexes (six or more units) can recycle the same materials as that of single-family residents in 32-, 64-, or 96-gallon bins.

SF Environment employees a Pay-As-You-Throw fee system (2012):

- $27.91/month for 32-gallon black bin
- Doubled for 64-gallon bin and tripled for 96-gallon black bin

Performance

Organics collected for composting:

- 600+ tons per day
- 160,000 tons annually

Multifamily:

- 95% of apartment buildings have composting service
- 99% of apartment buildings have recycling service

Overall:

- 80% MSW diverted from landfills

Figure 9.6 San Francisco's Organics Being Sent to Windrow Composting Facility. (From Jack Macy)

Contact:
Jack Macy
Jack.macy@sfgov.org

Los Angeles, California Recycling Program

Goals

The City developed a comprehensive recycling program in light of California's overall diversion goals of 25% by 1995 and 50% by 2000. The City initially implemented pilot programs using 14-gallon yellow (recyclables) and green (yard waste) using manual recycling vehicles. They have since progressed to 90-gallon wheeled carts due to the increase in volume using fully automated vehicles. The City has enacted a Zero Waste (90%) by 2025 goal.

Programs

The City manages a comprehensive program that annually collects over 240,000 tons of recyclables and 480,000 tons of yard trimmings.

- Residential—single-stream recycling using fully automated collection trucks with 90-gallon blue carts for recyclables and 90-gallon green yard waste carts.
- Multifamily (five or more unit) has mandatory recycling (July 2012). Multifamily units can enroll in the residential recycling program. Businesses can either deliver recyclables to drop-off centers or contract with permitted haulers who must report data to the City.

Performance

- Residential—979 tons per day of recyclables and 1783 tons per day of green waste
- 72% MSW recycled (2010)
- March 2009 issue of Waste and Recycling News identified LA as having highest recycling rate of 10 largest US cities

Contact:

Alex Helou
alex.helou@lacity.org

City of London, England

Goals

The Mayor of London, after extensive consulting, launched an ambitious waste management strategy by the end of 2011 with a vision of establishing London as a world class manager of its municipal waste.

Among the goals for London:

- To achieve zero municipal waste direct to landfill by 2025
- To reduce the amount of household waste produced to 790 kg per household by 2031 (equivalent to a 20% reduction)
- Recycle or compost at least 45% of municipal waste by 2015, 50% by 2020, and 60% by 2031

Current Programs

All 33 London boroughs offer at least a basic household curbside dry recycling collection service, although there is large disparity between the boroughs.

- All boroughs provide curbside collection services for paper, mixed cans, and plastic bottles. All except two boroughs collect glass at the curbside and all except one collect cardboard. Thirteen boroughs collect mixed plastics from curbside services.
- Nineteen boroughs provide a curbside commingled (mixed) recycling collection service. Twelve boroughs provide a curbside sort service, and two boroughs provide a mix of the two collection services.
- Seventeen boroughs collect dry recyclables in a box or wheelie bin. Eight boroughs use a sack, and eight boroughs use a combination of boxes, sacks, and wheelie bins. The color of recycling containers varies across boroughs.
- Twenty-six boroughs provide a weekly recycling collection service. Five boroughs provide a fortnightly recycling collection service. Two boroughs provide daily recycling collection services.
- All boroughs provide near entry (close to block or estate entrances) or bring site recycling banks for flats and estates, although there is great variation between boroughs on what materials are accepted.
- All except one borough provide a green-garden waste collection service. Eleven boroughs provide separate weekly curbside collections for food waste, and nine

boroughs collect food and green-garden waste together. Some boroughs provide food and green-garden waste collections for flats and estates.

Few boroughs have successfully tackled the problems of providing recycling and composting services to flats and other households that are not easily accessed from the street. Many boroughs continue to trial recycling and composting collection services in flats and estates, but some schemes have been withdrawn due to being too expensive, too difficult, and having low levels of participation.

Performance

- 32% MSW recycled (2010)

Figure 9.7 Informational kiosks and drop off containers in the downtown area.

Contact:
Lisa Moore
lisa.moore@london.gov.uk

Fort Hood Army Installation

Army Net Zero Vision

The Net Zero vision is a holistic approach to addressing energy, water, and waste at Army installations. It enables the Army to appropriately steward available resources, manage costs, and provide its Soldiers, Families, and Civilians with a sustainable future. The Net Zero vision ensures that sustainable practices will be instilled and managed throughout the appropriate levels of the

Army, while also maximizing operational capability, resource availability, and well-being.

Net Zero Waste Installation

On April 19, 2011, Fort Hood was selected as a Net Zero waste pilot installation to reach zero landfill by the year 2020. A Net Zero waste installation is an installation that reduces, reuses, and recovers waste streams, converting them to resource values with zero landfill over the course of a year. The components of Net Zero solid waste start with reducing the amount of waste generated, repurposing waste, maximizing recycling of waste stream to reclaim recyclable and compostable materials, recovery to generate energy as a byproduct of waste reduction, with disposal being nonexistent.

Performance

9700 tons in 2011
$1.8 million in 2011

Awards

2011 Secretary of the Army Environmental Quality Award recognized the recycling program as the largest successful, sustainable recycle facility in the Army by selling more than 17,000 tons of materials, and providing more than $500,000 to support Fort Hood community events and pollution prevention projects in 2010 and 2011.

Figure 9.8 Ft. Hood Recycle Facility. (From Michael Bush)

Contact:

Michael Bush, Operations Manager
Fort Hood Recycle
michael.k.bush9.naf@mail.mil

Bibliography

I consulted many books and magazine articles in my research, and I gratefully acknowledge the following works.

[1] Apotheker S. Reverse vending: a new market opportunity. Resource Recycling 1993; October:54–61.

[2] Apotheker S. Participation in drop-off recycling programs. Resource Recycling 1991; January:70–7.

[3] Arthur D. Little Inc. A report on advance deposit fees. Cambridge, MA; 1991.

[4] Artz NS, Robert Yoos E. Composting potential in municipal solid waste management. Prairie Village, Kansas: Franklin Associates, Ltd.; 1990.

[5] Baker S. Suddenly, there's aluminum everywhere. Bus Week 1993; October(25):46.

[6] Beaty EW. Litter got you stumped? Try clean community systems approach. : University of Tennessee: MTAS Publications, Tennessee Research and Creative Exchange, Knoxville; 1984.

[7] Bogert S, Jeffrey The economics of recycling. Resource Recycling 1993; Scptcmber:76–80.

[8] Broughton AC. MRFS go high tech. Recycling Today 1993; March:48–57.

[9] Broughton AC. State recycling laws: help or hindrance. Recycling Today 1993; February:121–30.

[10] Bullock D, Burk D. Commingled vs. curbside sort. BioCycle 1989; June:35–6.

[11] Cabaniss AD. Recycling education: the critical link. Resource Recycling 1993; February:62–5.

[12] California Assembly. Assembly bill 939; October 1991.

[13] Center For Plastics Recycling Research. Market research on plastics recycling, piscataway. Technical report #31. New Jersey: Rutgers, The State University of New Jersey; 1990.

[14] Chertow M. Garbage solutions: a public official's guide to recycling and alternative solid waste management technologies. Washington D.C.: National Resource Recovery Association and the United States Conference of Mayors; 1989.

[15] City of Irvine, California. What is zero waste, accessed at <http://www.cityofirvine.us/programs/zero-waste/what-is-zero-waste/>; 2013. Accessed March 3, 2013.

[16] Clarke MJ. The paradox and the promise of source reduction. Solid Waste and Power 1990; February:38–44.

[17] Cole MA, Leonas KK. Degradability of yard waste collection bags. BioCycle 1991; March:56–63.

[18] Commonwealth of Massachusetts A financial effect determination of mandatory recycling on Massachusetts cities and towns. Boston, MA: Office of the State Auditor; 1992.

[19] Conference Board of Canada. How Canada performs, accessed at <http://www.conferenceboard.ca/hcp/details/environment/municipal-waste-generation.aspx>; 2013. Accessed February 7, 2013.

[20] Congress of the United States. Office of technology assessment, facing America's trash: what next for municipal solid waste. Washington, D.C.; 1989.

[21] Connecticut Department of Energy and Environmental Protection. Using social media to promote recycling, accessed at <http://www.ct.gov/dEep/cwp/view.asp?a=2714&q=487970&deepNav_GID=1645>; 2013. Accessed January 7, 2013.

[22] Couling T. The pros and cons of composting. American City and County 1990; December:56–62.

[23] City of San Jose. Zero waste strategic plan. Environmental Services Department; November 2008.

[24] Crampton N. Full cost accounting: what is it? Will it help or hurt recycling. Resource Recycling 1993; September:57–61.

[25] DSM Environmental Services. Recycling economic information study update: Delaware, Maine, Massachusetts, New York, and Pennsylvania. Prepared for the Northeast Recycling Council; 2009.

[26] Environmental Defense Fund. Coming full circle—successful recycling today. New York, NY: EDF; 1988.

[27] Fishbein BK, Gelb C. Making less garbage—a planning guide for communities. New York, NY: INFORM; 1992.

[28] Flavell JW. A tribute to piaget. Society for research in child development newsletter. Chicago, IL: The Society for Research in Child Development; 1980.

[29] Garino RJ. UBC recycling on a roll, scrap processing and recycling; May/June 1991. 115–117.

[30] Gatton D. Municipal waste management association. Interview; 1993.

[31] Geller ES, Lehman GR. Motivating desirable waste management behavior: applications of behavior analysis. J Resour Manage Technol 1986; December:65–7.

[32] Glenn J. An industry shapes up for recycled plastics. BioCycle 1991; January:38–67.

[33] Goldman M. What do those recycling numbers mean. The Weston Way; 1990.

[34] Grove C. Demystifying the curbside plan. BioCycle 1989; June:45–7.

[35] Haverty TE, Regan R. Recycling: equal access to capital. Resource Recycling 1993; October:75–80.

[36] HDR Engineering, Inc. Muscatine county conceptual solid waste management and recycling system. Prepared for Muscatine. Iowa; September 1993.

[37] HDR Engineering, Inc. Public opinion survey findings: Monmouth County, New Jersey. Prepared for Municipal Solid Waste Management Program; May 1993.

[38] HDR Engineering, Inc. Yard waste collection and composting options. Prepared for the El Dorado Hills Community Service District; October 1992.

[39] HDR Engineering, Inc. Union county recycling plan update study. Prepared for Union County Utilities Authority; June 1991. 61–149.

[40] HDR Engineering, Inc. Technology overview and evaluation. Prepared for the South Plains Association of Governments; December 1991.

[41] HDR Engineering, Inc. Pilot yard waste collection study. Prepared for Orange County, Florida, Orlando; September 1991.

[42] HDR Engineering, Inc. High-rise and multi-family recycling feasibility study. Prepared for the Northern Virginia Planning District Commission; July 1990.

[43] Hickman LH. Solid waste association of North America. Interview; December 1993.

[44] Hickman HL Jr, Reimers EG. Unit pricing for municipal solid waste management services. International directory of solid waste management. 1993/4 the IWSA yearbook; November 1993. 22–22.

[45] Hickman Jr. HL. Looking at municipal solid waste composition and mandatory removal rates in North America. Silver Spring, MD: Solid Waste Association of North America; 1991.

[46] Hunsicker M. The rest of the story. Recycling Today 1992; July:48–54.
[47] International Solid Waste Association. Alternative waste conversion technologies. White paper; January 2013.
[48] Kelly S. Large scale yard waste composting. BioCycle 1993; September:30–2.
[49] Kovacs WL, Pellegrini ME. Flow control: the continuing conflict between free competition and monopoly public service. Washington, D.C.: Resource Recovery Report; 1993.
[50] Kraten S. Market failure and the economics of recycling. Environmental Decisions 1990; April:20–5.
[51] Kropf B, Mixon C. Drop-off recycling in tampa: maximizing efficiency and quality. Resource Recycling 1993; January:51–6.
[52] Littleton G. Recycling for success. Envirosouth 1990; Spring:15–17.
[53] Los Angeles County Solid Waste Committee/Integrated Waste Management Committee. Task force adopts key definitions and new solid waste paradigm. MSW management; October 31, 2012.
[54] McEntee K. Paper-makers pushed to do more recycling. Recycling Today 1990; February:120–6.
[55] McKinstry RB, Prendergast WM. Drafting a mandatory recycling ordinance. BioCycle 1989; June:78–80.
[56] Mersky RL. Designing a recycling program. Public Works 1989; May:67–9.
[57] Metro Vancouver. Integrated solid waste and resource management for the greater vancouver regional district and member municipalities. Metro Vancouver; 2010.
[58] Mishkin A. Beveridge Diamond PC. Statement quoted in the minutes of the recycling forum; January 1992.
[59] Monmouth County. Monmouth County public opinion survey. New Jersey; May 1993.
[60] Murdoch JD, Collins RJ, Williams JD. Mandatory recycling—five years later. MSW Management 1992; November:4449.
[61] National Recycling Coalition, Inc. State of recycling—1993. Washington, D.C.; August 1993.
[62] National Recycling Coalition, Inc. State of recycling—1993. Washington, D.C.; October 1993.
[63] National Research Council America's climate choices: panel on advancing the science of climate change; advancing the science of climate change. Washington, D.C.: The National Academies Press; 2010.
[64] National Solid Waste Management Association. The cost to recycle at a materials recovery facility. Washington, D.C.; 1992.
[65] New York State. Beyond waste: a sustainable material management strategy. NY: Department of environmental conservation; 2010.
[66] O'Brien JK, Williams JF. Calculating your maximum recycling potential. Environmental Decisions 1990; December:19–21.
[67] O'Brien JK. Integrated collection: the key to economical curbside recycling. Solid Waste and Power Magazine 1991; August:62–70.
[68] Oneida-Herkimer Solid Waste Authority. Local solid waste management plan. Utica, NY; 1990.
[69] Oregon Environmental Quality Commission. Materials management in Oregon—2050 vision and framework for action. State of Oregon Department of Environmental Quality; 2012.
[70] Wayne P, Pferdehirt PE. AICP, planning bigger, faster, more flexible MRFs. Solid Waste and Power 1990; October:53–60.
[71] Phipps PH. Scrap aluminum lowers sheet raw material costs. Resource Recycling 1992; March:15–16.

[72] Plastics Recycling Foundation. Plastics recycling: an overview. Washington, D.C.; 1990.

[73] Plastics Recycling Foundation. Plastics recycling: from vision to reality. Washington, D.C.; 1990.

[74] Platt B, Zachary J. Co-collection of recyclables and mixed waste: problems and opportunities. Washington, D.C.: Institute For Local Self-Reliance; September 1991.

[75] Porter J, Winston DR. Status report on municipal solid waste recycling, the solid waste task force. Government Programs Steering Committee of the American Institute of Chemical Engineers; July 1992.

[76] Powell J. The ups and downs in bottle to bottle plastics recycling. Resource Recycling 1992; May

[77] Powell J. Recycling and the law: the flow control battle. Resource Recycling 1993; September:35–8.

[78] Powell J. The decline of the legislator, the rise of the regulator: recent trends in state recycling programs. Resource Recycling 1992; October:44–5.

[79] Powell J. The common recyclable: the growth in plastic bottle recovery. Resource Recycling 1991; January:40–6.

[80] Powers R. Keep America Beautiful Inc. Interview; December 1993.

[81] Rapkin A. Aluminum cans: market dynamics. Resource Recycling 1993; October:32–9.

[82] Ragsdale JV, Stasis P, Rudd MJ, Bradshaw J. Mulch production from yard trash. BioCycle 1992; September:34–7.

[83] Resource Integration Systems, Ltd. Public sector commercial recycling programs: 10 case studies. Prepared for the City of Los Angeles, Office of Integrated Waste Management; June 1990.

[84] Rhea M. National recycling coalition, Inc. Interview; December 1993.

[85] Rogers F. Mister rogers activities for young children about the environment and recycling. Pittsburgh, PA: Family Communications, Inc.; 1990.

[86] Rogoff MJ, Williams JF. The waste hurdles. American City and County 1993; January:46–50.

[87] Ruston J. Developing recycling markets for the components of residential mixed paper. Resource Recycling 1992; January

[88] Santrock JW, Yussen SR. Child development and introduction,. Dubuque, IA: William C. Brown Publishers; 1989. pp. 13, 149, 168–169, 298.

[89] Scarlett L. Recycling costs: clearing away some smoke. Solid Waste and Power 1993; August:12–17.

[90] Schickendanz JA, Hansen K, Forsyth PD. Understanding children. Mountain View, CA: Mayfield Publishing Company; 1990. 60–68.

[91] SCS Engineers, Inc. Final report: Kent County, Delaware waste composition study. Prepared for the Delaware Solid Waste Authority; August 1990.

[92] SCS Engineers, Inc. First seasonal report, waste composition study at the northern solid waste management center, New Castle, Delaware. Prepared for the Delaware Solid Waste Authority; November 1990.

[93] Selby MD. Yard waste collection. BioCycle 1989; June:52–4.

[94] Sherman SP. Trashing a $150 billion business. Fortune; August 28, 1989, 92–98.

[95] Skinner BF. The behavior of organisms: an experimental analysis. New York, NY: Appleton-Century; 1938.

[96] Slovin J. Developing markets to close the loop. World Waste 1993; October:42–6.

[97] Solid Waste Association of North America. Policy positions on municipal solid waste management. Silver Spring, MD; December 1993, 15–16.

[98] Solid Waste Association of North America. Manager of landfill operations training course manual. Silver Spring, MD; May 1992.

[99] Solid Waste Association of North America. Compendium of solid waste management terms and definitions. Silver Spring, MD; 1991. Publication #GR-G 0001.

[100] Solid Waste Authority of Palm Beach County. Prepared for Florida, Palm Beach County Recycling Program. West Palm Beach, FL; October 1989.

[101] Sound Resource Management Group, Inc. The economics of recycling and recycled materials. Prepared for the Clean Washington Center. Seattle, Washington, D.C.; June 1993.

[102] Trombly J. Developing non-traditional glass markets. Resource Recycling 1991; October

[103] U.S. Conference of Mayors. U.S. Solid waste composting facility profiles. Washington, D.C.; May 1992.

[104] U.S. Conference of Mayors. Recycling in America: profiles of the nation's resourceful cities. Washington, D.C.; June 1991.

[105] U.S. Environmental Protection Agency. Municipal solid waste generation, recycling and disposal in the United States: facts and figures for 2010, <http://www.epa.gov/wastes/nonhaz/municipal/pubs/msw_2010_rev_factsheet.pdf>; 2011. Accessed October 3, 2012.

[106] U.S. Environmental Protection Agency. Beyond RCRA: waste and materials management in the year 2010. U.S. Environmental Protection Agency; 2002.

[107] U.S. Environmental Protection Agency. Decision-makers guide to solid waste management, Washington, D.C.: EPA/530-SW-89-072; November 1990.

[108] U.S. Environmental Protection Agency. Variable rates in solid waste: handbook for solid waste officials, vol. 1—executive summary, EPA/530-SW-90-084A; September 1990.

[109] U.S. Environmental Protection Agency. Charging households for waste collection and disposals: the effects of weight or volume-based pricing on solid waste management. Washington, D.C.: EPA/530-SW-90-047; 1990.

[110] Vermont Agency of Natural Resources. Materials management in vermont: history of materials management and planning update. Vermont Department of Environmental Conservation; 2010.

[111] Waste Management, Inc. Recycling in the 90's, Oak Brook, IL: WM; 1992.

[112] Watson T. Reverse vending revisited: great losses and high hopes. Resource Recycling 1990; March:26–77.

[113] Weitz KA, et al. The impact of municipal solid waste management on greenhouse gas emissions in the United States. Air and Waste Management Association, Pittsburgh, PA. 2002;52:1000–11.

[114] Williams J, Weitz KA, Thorneloe SA, Nishtala SR, Yarkosky S, and Zannes M. Beginning with preschoolers: a recycling education initiative, air & waste management association. Proceedings of the Annual Meeting, Pittsburgh, PA; June 1991.

[115] Williams J. Mega-MRFs & full service contracts: a key to public/private ventures. MSW Management 1992; October

[116] Williams J. Municipal solid waste composting: looking after the community's interests, air & waste management association. Proceedings of the Annual Meeting; June 1992.

[117] Williams J, Kulik A. American's attitudes about recycling. American City and County 1992; July:48–50.

[118] Wood JJ. The challenge of multi-family recycling. Resource Recycling 1991; June:34–40.

[119] World Business Council for Sustainable Development. Sustainable consumption facts and trends from a business perspective; 2008.

[120] Yergin D. The prize: the epic quest for oil, money and power. New York, NY: Simon and Shuster; 1992.

[99] Solid Waste Association of North America. Compendium of solid waste management terms and definitions. Silver Spring, MD; 1991. Publication #GR-G 0001.
[100] Solid Waste Authority of Palm Beach County. Prepared for Florida; Palm Beach County Recycling Position. West Palm Beach, FL; October 1989.
[101] Sound Resource Management Group, Inc. The economics of recycling and recycled materials. Prepared for the Clean Washington Center. Seattle, Washington, D.C.; June 1993.
[102] Truini, J. Developing nontraditional glass markets. Resource Recycling 1991; October.
[103] U.S. Conference of Mayors. U.S. solid waste composting facility profiles. Washington, D.C.; May 1989.
[104] U.S. Conference of Mayors. Recycling in America: profiles of the nation's resourceful cities. Washington, D.C.; June 1991.
[105] U.S. Environmental Protection Agency. Municipal solid waste generation, recycling, and disposal in the United States: facts and figures for 2010. <http://www.epa.gov/wastes/nonhaz/municipal/pubs/msw_2010_factsheet.pdf>; 2011 [Accessed October 2013].
Management of MSW — executive summary. EPA/530-SW-90-042A. September 1990.
[108] U.S. Environmental Protection Agency. Charging households for waste collection and disposal: the effects of weight or volume-based pricing on solid waste management. Washington, D.C.: EPA/530-SW-90-047; 1990.
[111] Watson T. Recycling in the 90s. Resource Recycling 1991.
[112] Watson T. Recycling revisited: glass, metals, and other materials. Resource Recycling 1990; March:26–74.
[113] Weitz KA, et al. The impact of municipal solid waste management on greenhouse gas emissions in the United States. Air and Waste Management Association. Pittsburgh, PA; 2002;52:1000–11.
[114] Williams J, Weitz KA, Thorneloe SA, Nishtala SR, Yarkosky S, and Zannes M. Beginning with preschoolers: a recycling education initiative in a waste management association. Proceedings of the Annual Meeting. Pittsburgh, PA; June 1991.
[115] Williams J, Miller M, Myers A. Solid waste conference: a key to municipal solid waste management. 1992; October.
[116] Williams L. Applied solid waste composting: making sense of the community's materials. Solid waste management case in the Florida meeting. June 1992.
[117] Williams R, Smith D. American attitudes about recycling. American City and County 1992;107:10–12.
[118] Wirth J. The challenge of multi-family recycling. Resource Recycling 1991; June:1.
[119] World Business Council for Sustainable Development. Sustainable consumption facts and trends from a business perspective. 2008.
[120] Yergin D. The prize: the epic quest for oil, money, and power. New York, NY: Simon and Schuster; 1992.

Glossary

Aeration The process of exposing bulk material, such as compost, to air.

Aerobic A biochemical process or condition occurring in the presence of oxygen.

Anaerobic A biochemical process or condition occurring in the absence of oxygen.

Anaerobic Digestion The controlled biological breakdown of biodegradable organic matter in the absence of oxygen.

Baler A machine used to compress recyclables into bundles to reduce volume.

Biodegradable Materials Waste material which is capable of being broken down by microorganisms into simple, stable compounds such as carbon dioxide and water. Most organic wastes such as food remains and paper are biodegradable.

Biogas A gas produced through anaerobic digestion and is primarily composed of methane and carbon dioxide, but also may contain impurities such as hydrogen sulfide.

Biomass Amount of living matter in the environment.

British Thermal Unit This is a unit of measure for the amount of energy a given material contains as energy is released as heat during the combustion. One Btu is the quantity of heat required to raise the temperature of one pound of water by one degree Fahrenheit.

Buy Back Center A facility where individuals bring recyclables for payment.

Biosolids Solids derived from primary, secondary, or advanced treatment of domestic wastewater which has been treated through one or more controlled processes that significantly reduce pathogens and reduce volatile solids or chemically stabilize solids to the extent that they do not attract vectors.

Capital Costs Those direct costs incurred to acquire real property assets such as land, buildings, machinery, and equipment.

Collection Routes Established routes followed in the collection of refuse and recyclables from homes, businesses, commercial plants, and other locations.

Collection System The total process of collecting and transporting solid waste. It includes storage containers, collection crews, vehicles, equipment, and management and operating procedures.

Commercial Waste Waste materials originating in wholesale, retail, or service establishments such as office buildings, stores, markets, theaters, hotels, and warehouses.

Compost A relatively stable decomposed organic material; the result of the composting process.

Composting The controlled biological decomposition of organic solid waste under aerobic conditions.

Composting Facility A site or facility composting feedstocks to produce a useful product through a managed process of controlled biological decomposition. Examples of composting facilities include sites used for composting windrows and piles, anaerobic digestion, vermiculture, vermicomposting, and agricultural composting.

Construction and Demolition Waste (C&D) These are waste building materials, packaging, grubbing waste and rubble resulting from construction, dredging materials, remodeling, and demolition operations on pavements, houses, commercial buildings, and other structures. These kinds of materials usually include used lumber, metal parts, packaging materials, boxes, sheet metal, and other materials.

Conversion Technology The use of chemical or thermal processes to convert solid waste to fuels or other similar useful products. These technologies include the following: pyrolysis, gasification, anaerobic digestion, hydrolysis, and distillation.

Conversion Technology Facility A facility that uses primarily chemical or thermal processes (changing from solid to liquid through heating without changing chemical composition) to produce fuels, chemicals, or other useful produces from solid waste. These chemical or thermal processes include, but are not limited to, distillation, gasification, hydrolysis, pyrolysis, thermal depolymerization, transesterification, and animal rendering, but do not include direct combustion, composting, anaerobic digestion, melting, or mechanical recycling. Mills that primarily use mechanical recycling or melting to recycle materials back into similar materials are not considered to be conversion technology facilities, even if they use some chemical or thermal processes in the recycling process.

Cost Everything given up to acquire a material or service, or achieve a goal.

Cost Savings The monetary savings realized through waste reduction and recycling as a result of avoiding landfill or other disposal processes; sometimes referred to as "avoided cost."

Cullet Clean, generally color sorted, crushed glass used to make new glass products.

Curbside Collection Collection of recyclable materials at the curb, often from special containers, to be taken to various processing facilities.

Digestate Both solid and liquid substances that remain following anaerobic digestion of organic material in a composting facility. "Solid digestate" means the solids resulting from anaerobic digestion, and "liquid digestate" means the liquids resulting from anaerobic digestion.

Dirty Materials Recovery Facilities A facility that accepts a mixed waste stream and then proceeds to separate our designated recyclable materials through a combination of manual and mechanical sorting.

Disposal These are activities associated with the long-term handling of solid waste that are collected and are no further use.

Diversion Rate A measure of the amount of waste material being diverted for recycling compared with the total amount that was previously thrown away.

Downstream Those actions and impacts that occur after that point in the life cycle, at any point on a product's life cycle.

Drop-off Center A method of collecting recyclable or compostable materials in which the materials are taken by individuals to collection sites and deposited into designated containers.

Dual-Stream Recycling Recycling processes in which the waste streams are separated at a materials recovery facility. One stream is usually fiber and the other containers.

End-of-Life The point at which a product or material is no longer useful to the person possessing it and is either discarded or abandoned.

Energy Recovery Recovery in which all or a part of the solid waste materials are processed to use the heat content, or other forms of energy, of or from the material. Energy recovery includes the direct combustion of solid waste in an energy recovery facility and the production of fuels intended to be burned as an energy source, such as the pyrolysis of plastics to produce fuel oils or the grinding of wood waste to produce combustion fuel.

Energy Recovery Facility A facility that directly combusts solid waste and uses the heat energy generated for some useful purpose such as to produce electricity or produce steam to be used in an industrial process.

Enterprise Fund A fund for a specific purpose that is self-supporting from the revenue it generates.

Extended Producer Responsibility (EPR) A mandatory type of product stewardship that includes, at a minimum, the requirement that the producer's responsibility for the product extends to post-consumer management of that product and its packaging.

E-Waste A computer, computer monitor, and computer peripheral device containing a cathode ray tube, printer, or television.

Ferrous Metals Pertaining to, or derived from, iron; often used to refer to materials that can be removed from the waste stream by magnetic separation.

Financial Assurance A plan for setting aside financial resources or otherwise assuring that adequate funds are available to properly close and to maintain a monitor at a disposal site after the site is closed according to the requirements of a permit issued by the department.

Flow Control A legal or economic means by which waste is directed to particular destinations.

Food Waste Animal or vegetable wastes resulting from the handling, storage, sale, preparation, cooking, and serving of foods.

Front-End System A process for salvaging certain reusable materials from the waste before combustion or other processing.

Front-End Loader A collection vehicle with arms that engage a detachable container, move it up over the cab, empty it into the vehicle's body, and return it to the ground.

Front-End Recovery The mechanical processing of discarded solid wastes into separate constituents.

Garbage Spoiled or waste food that is thrown away, generally defined as wet food waste; although in common usage, garbage refers to all materials that are discarded as unnecessary.

Generation The act or process of producing solid waste.

Glass An inorganic product of fusion that has cooled to a rigid condition without crystallizing.

Grade A term applied to a paper or pulp which is ranked on the basis of its use, appearance, quality, manufacturing history, raw materials, performance, or a combination of these factors.

Greenwashing The practice of making an unsubstantiated or misleading claim about the environmental benefits of a product, service, or technology.

Green Waste A combination of nonanimal food and yard waste collected and composted together.

Hammermill A type of crusher used to break up waste materials into smaller pieces or particles, which operates by using rotating and failing heavy hammers.

Hazardous Waste Discarded, useless, or unwanted materials or residues and other wastes are defined as hazardous waste.

HDPE High-density polyethylene, a plastic resin used to make items such as plastic milk and detergent containers, and base cups for plastic soft drinks.

High-Grade Paper Relatively valuable types of paper such as computer printout, white ledger, and tab cards.

Household Hazardous Waste Any waste from households, hotels or motels, bunkhouses, ranger stations, crew quarters, camp grounds, picnic grounds, and day-use recreation areas that would be subject to regulation as hazardous wastes if it were not from households.

Incinerator Any device used for the reduction of combustible solid wastes by burning under conditions of controlled airflow and temperature.

Industrial Waste Those waste materials generally discarded from industrial operations or derived from manufacturing processes.

Infectious Waste Biological waste, cultures and stocks, pathological waste, and sharps.

Institutional Waste Solid wastes generated by schools, hospitals, universities, museums, governments, and other institutions. Some communities define institutional solid waste as commercial solid waste.

Integrated Solid Waste Management A practice of disposing of solid waste using several complementary components, such as waste reduction, recycling, composting, energy recovery, and landfilling.

In-Vessel Composting A composting method in which the compost is continuously and mechanically mixed and aerated in a large, contained area.

Intermediate Processing Center A type of materials recovery facility (MRF) that processes residentially collected mixed recyclables into new products available for market; often used interchangeably with MRF. An acronym is IPC.

Investment Tax Credit A reduction in taxes permitted for the purchase and installation of specific types of equipment and other investments.

Junk Old or scrap metals, rope, rags, batteries, paper, rubber, junked, dismantled or wrecked automobiles or parts thereof which are not held for sale for remelting purposes; unprocessed materials suitable for reuse or recycling, commonly referred to as secondary materials.

Kraft Paper A paper made predominantly from wood pulp produced by a modified sulfate pulping process. It is a comparatively coarse paper particularly noted for its strength; in unbleached grades is used primarily as a wrapper or packaging material.

Landfill A facility for the disposal of solid waste involving the placement of solid waste on or beneath the land surface.

Leachate Liquid that has come into direct contact with solid waste and contains dissolved, miscible, and/or suspended contaminants as a result of such contact.

Life Cycle Assessment (LCA) A standardized process used to estimate the impact that a product or process has over the whole of its life span, including extraction of raw materials, production, transport, use, and disposal.

Magazine Paper A variety of coated and uncoated papers used in magazines and similar periodicals.

Magnetic Separation A system used to remove ferrous metals from other materials through the use of magnets.

Mandatory Recycling Programs requiring by ordinance or statute that residents or businesses keep specific materials from their solid wastes.

Manila Paper Indicates color and finish and not the use of manila hemp.

Manual Separation The separation of recyclable materials from waste by hand sorting.

Materials Management An approach to reduce environmental impacts by managing through all stages of their life. Materials management identifies impacts and actions across the full cycle of materials and products as they move through the economy—from raw material extraction to product design and manufacture, transport, consumption, use, reuse, recycling, and disposal.

Materials Recovery The concept of resource recovery, emphasis is on separating and processing waste materials for beneficial use or reuse.

Materials Recovery Facility A common acronym is MRF.

Maximum Recycling Potential The maximum amount of recycling possible for a community given an ideal market, regulatory, citizen participation, and technological limits. MRP is sometimes used as an acronym.

Mechanical Separation The separation of waste into various components using mechanical means, such as cyclones, trommels, and screens.

Mixed Kraft Bags Consists of baled used kraft bags free from twisted or woven stock and other similar objectionable materials.

MSW Composting Mixed or municipal solid waste composting, the controlled degradation of municipal solid waste including some form of presorting to remove noncompostable inorganic materials.

Mulch Ground or mixed yard wastes placed around plants to prevent evaporation of moisture and freezing of roots and to nourish the soil.

Municipal Collection The collection of solid wastes by a public agency.

Municipal Solid Waste Includes nonhazardous waste generated in households, commercial and business establishments, institutions, and light industrial wastes; it excludes industrial process wastes, agricultural wastes, mining wastes, and sewage sludge.

Newsprint A generic term used to describe paper of the type generally used in the publication of newspapers.

NIMBY Acronym for “Not in My Backyard”; expression of opposition to the siting of a facility based on the particular location proposed.

Nonferrous Metal Metals which contain no iron, such as aluminum, copper, brass, and bronze.

Organic Waste Waste material from substances composed primarily of chemical compounds of carbon in combination with other elements, primarily hydrogen. These materials include paper, wood, food wastes, plastics, and yard waste.

Packer Truck Type of solid waste collection vehicle used for residential collection that compacts refuse into high-density masses for maximum collection efficiency. It can incorporate a rear loading or top loading device.

Paper The name for all kinds of matted or felted sheets of fiber formed on a fine screen for a water suspension.

Participation Rate A measure of the number of people participating in a recycling program compared to the total eligible.

Pathogen An organism capable of producing disease.

Per Capita Disposal Rate The average amount of waste disposed (landfill or incinerated) per person per year for a given year.

PET Polyethylene terephthalate, a plastic resin used to make packaging, commonly used to make plastic soft drink bottles.

Plastics Nonmetallic compounds that result from a chemical reaction, and are molded or formed into rigid or pliable construction materials and fabrics.

Polypropylene A heavy-duty plastic.

Polystyrene A hard, dimensionally stable thermoplastic that is easily molded.

Polyvinyl Chloride A common plastic material which is tasteless, odorless, and generally insoluble; acronym is PVC.

Postconsumer Recycling The reuse of materials generated from residential and commercial waste, excluding recycling of material from industrial processes that has not reached the consumer, such as glass broken in the manufacturing process.

Price Preference A means by which an incentive is provided to purchase recycled goods even if they are more expensive than nonrecycled goods.

Processing of Wastes Any technology designed to change the physical form or chemical content of solid waste including, but not limited to, baling, composting, classifying, hydropulping, incinerating, and shredding.

Pulping The operation of reducing a cellulosic raw material into a pulp suitable for further processing into paper.

Putrescible Waste Solid waste containing organic material that can be rapidly decomposed by microorganisms, and which may give rise to foul smelling, offensive products during such decomposition, or which is capable of attracting or providing food for bids and potential disease vectors such as rodents and flies.

Recycled Material Material that can be utilized in place of a raw or virgin material in manufacturing a product and consists of materials derived from postconsumer waste, industrial scrap, material derived from agricultural wastes or other items, all of which can be used in the manufacture of new products.

Recycling Specifically separating a given waste material from the waste stream and processing it so that it may be used again as a raw material for products which may or may not be similar to the original.

Recycling Center A place where people bring items to be recycled.

Recycling Rate The quantity of material recycled compared to the sum of recycled and disposed material.

Residential Waste Waste materials generated in single and multiple family homes; when multiple family units exceed four, these wastes are usually collected in large containers by commercial haulers.

Residue Materials remaining after processing, composting, and recycling.

Resource Recovery The process of obtaining useful material or energy from solid waste and includes energy recovery, material recovery, and recycling.

Reuse The return of a commodity into the economic stream for use in the same kind of application as before without change in its identity.

Roll-off Container A steel box with wheels used to collect waste at a site, such as a construction site, that can be rolled onto a truck using a winch and then taken to a disposal facility for discharge. The empty container can then be trucked to another site and rolled off the truck for stationary waste collection.

Rubber A natural or synthetic elastic material comprised of polymers. Chemical treatment can enhance properties required for tires, shoes, insulation, and other products.

Rubbish A nonputrescible solid wastes, including ashes, consisting of both combustible and noncombustible materials, such as paper, cardboard, tin cans, wood, glass, bedding, crockery, or litter of any kind.

Salvage The controlled removal of reusable, recyclable, or otherwise recoverable materials from solid wastes at a solid waste disposal site.

Scalehouse A building located at the entrance of a recycling or disposal facility where weigh scales is placed.

Scavenger One who illegally removes materials at any point in the solid waste system. It is an alternate name for waste hauler or carter.

Scrap Discarded or rejected industrial waste material suitable for reprocessing.

Secondary Materials Materials that are used in place of a primary or raw material in manufacturing a product, often handled by dealers and brokers in "secondary markets."

Shredder A size reduction machine which tears or grinds materials to a smaller and more uniform particle size. Shredding processes are also called size reduction, grinding, milling, comminution, pulverization, hogging, granulating, breaking, macerating, chipping, crushing, cutting, and rasping.

Single-Stream Recycling A recycling process where the recycling stream is fully commingled in one bin for collection curbside. These materials are then transported to a single-stream MRF.

Small Quantity Generator A generator that produces less than 100kg of hazardous waste per month (or accumulates less than 100kg at any one time) or one who produces less than 1 kg of acutely hazardous waste per month (or accumulates less than 1 kg of acutely hazardous waste at any one time).

Solid Waste A general term for discarded materials destined for disposal, but not discharged to a sewer or the atmosphere. Solid waste(s) can be composed of a single material or a heterogeneous mix of various materials including semisolids.

Solid Waste Disposal The disposal of all solid wastes through landfilling, incineration, composting, chemical treatment, and any other method which prepares solid wastes for final disposition.

Solid Waste Management A planned program for effectively controlling the generation, storage, collection, transportation, processing and reuse, conversion or disposal of solid waste in a safe, sanitary, aesthetically acceptable, environmentally sound and economic manner. It includes all administrative, financial, environmental, legal, and planning functions, as well as the operational aspects of solid waste handling and resource recovery systems.

Source Reduction The design, manufacture, acquisition, and reuse of materials including products and packaging, so as to minimize the quantity and/or toxicity of waste produced. Source reduction prevents waste either by redesigning products or by otherwise changing societal patterns of consumption, use, and waste generation.

Source Separation The segregation of specific waste materials at the point of discard for separate collection.

Special Wastes Hazardous wastes by reason of their pathological, explosive, radioactive, or toxic nature.

Specification Clear and accurate description of the technical requirement for materials, products, or services. It specifies the minimum requirement for quality and construction of materials and equipment necessary for an acceptable product in the form of written descriptions, drawings, commercial designations, industry standards, and other descriptive references.

Static Pile System A windrow composting method in which air ducts are generally installed under or in the base of compost piles so air can be blown into or drawn through the pile

Subtitle D That portion of RCRA dealing with nonhazardous solid waste treatment, storage, and disposal facilities.

Sustainability Using, developing, and protecting resources in a manner that enables people to meet current needs and provides that future generations can also meet future needs, from the joint perspective of environmental, economic, and community objectives.

Thermophiles Bacteria or other microorganisms which grow best at temperatures of roughly 45–60°C.

Tin Can Made from tin-plated steel.

Tipping Fee The charge to unload waste materials at a transfer station, processing plant, landfill, or other disposal site.

Tipping Floor Unloading area for vehicles that are delivering waste materials to transfer stations, incinerators, or other processing plants.

Tons per Day Usually refers to the capacity of a paper mill, refuse processing plant, landfill, etc.

Transfer Station Supplemental transportation systems, an adjunct to route collection vehicles to reduce haul costs or add flexibility to the operation. Typically route vehicles empty into a large hopper from which large semitrailers, railroad gondolas, or barges are filled. There may be some compaction of refuse. Transfer stations may be fixed or mobile, since the larger compacting collection vehicles can serve this function.

Transfer Trailer A vehicle used to transport large quantities of waste over long distances.

Trash Material considered worthless, unnecessary, or offensive that is usually thrown away; generally defined as dry waste material, but in common usage, it is a synonym for garbage, rubbish, or refuse.

Trommel A large revolving cylindrical screen used as a waste separation technique.

Tub Grinder Machine used to grind or chip wood wastes for mulching, composting, or size reduction.

Variable Can Rate A charge for solid waste services based on the volume of waste generated measured by the number of containers set out for collection.

Vegetative Wastes Plant clippings, prunings, and other discarded material from yards and gardens. Also known as yard rubbish or yard waste.

Vermicomposting The controlled and managed process by which live worms convert solid waste into dark, fertile, granular excrement.

Vermiculture The raising of earth worms for the purpose of collecting castings for composting or enhancement of a growing medium.

Volume Reduction The processing of waste materials so as to decrease the amount of space the materials occupy, usually by compacting or shredding (mechanical), incineration (thermal), or composting (biological).

Waste Useless, unwanted, or discarded material resulting from natural community activities. Wastes include solids, liquids, and gases. Solid wastes are classified as refuse.

Waste Exchange A computer and catalog network to redirect waste materials back into the manufacturing process by matching companies generating specific wastes with companies that use those wastes as manufacturing inputs.

Waste Paper It refers to any paper or paper product which has lost its value for its original purpose and has been discarded. The term is most commonly used to designate paper suitable for recycling, as paper stock.

Waste Processing An operation such as shredding, compaction, composting, or incineration, in which the physical or chemical properties of wastes are changed.

Waste Reduction The practice of producing smaller quantities of disposable waste. Waste reduction usually entails changing manufacturing processes and packaging practices to foster more recycling and less dependency on disposable goods.

Waste Stream The waste output of a region, community, or facility.

White Goods Discarded kitchen and other large, enameled appliances, as washing machines and refrigerators.

White Paper Printers term of unprinted paper, even if colored.

Windrow Composting material stacked in a triangular prism shape.

Windrow System A composting technique in which waste is placed in either aerated static piles or turned, windrowed piles to digest.

Wood Fiber Elongated, thick-walled cells of wood, commonly called "fiber."

Wood Pulp A fibrous raw material derived from wood for use in most types of paper manufactured by mechanical or chemical means both from hardwood and softwood trees.

Yard Waste Leaves, grass clippings, prunings, and other natural organic matter discarded from yards and gardens.

Index

Note: Page numbers followed by "*f*", "*t*" and "*b*" refers to figures, tables and boxes respectively.

R

S

Y

Z

Printed and bound by CPI Group (UK) Ltd, Croydon, CR0 4YY
08/05/2025
01864838-0001